Math Mammoth
Grade 3-B Worktext

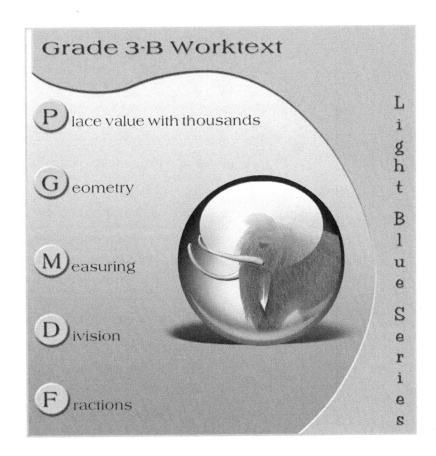

By Maria Miller

Copyright 2017 - 2018 Maria Miller
ISBN 978-1481012584

Edition 1/2018

All rights reserved. No part of this book may be reproduced or transmitted in any form or by any means, electronic or mechanical, or by any information storage and retrieval system, without permission in writing from the author.

Copying permission: Permission IS granted to reproduce this material to be used with one (1) teacher's students by virtue of the purchase of this book. In other words, one (1) teacher MAY make copies of these worksheets to be used with his/her students. Permission is not given to reproduce the material for resale. Making the file(s) available on any website for the purpose of sharing is strictly prohibited. If you have other needs, such as licensing for a school or tutoring center, contact the author at http://www.MathMammoth.com/contact.php.

Contents

Foreword	5

Chapter 6: Place Value with Thousands

Introduction	6
Thousands	8
Four-Digit Numbers and Place Value	12
Which Number Is Greater?	16
Mental Adding and Subtracting	18
Add 4-Digit Numbers with Regrouping	22
Subtract 4-Digit Numbers with Regrouping	24
Rounding to the Nearest Hundred	28
Estimating	31
Word Problems	34
Mixed Review Chapter 6	37
Review Chapter 6	39

Chapter 7: Geometry

Introduction	41
Shapes	46
Some Special Quadrilaterals	50
Perimeter	53
Problems with Perimeter	56
Getting Started with Area	59
More about Area	61
Multiplying by Whole Tens	65
Area Units and Problems	69
Area and Perimeter Problems	73
More Area and Perimeter Problems	75
Solids	78
Mixed Review Chapter 7	80
Geometry Review	82

Chapter 8: Measuring

Introduction	84
Measuring to the Nearest Fourth-Inch	87
Centimeters and Millimeters	91
Line Plots and More Measuring	95
Feet, Yards, and Miles	98
Meters and Kilometers	100
Pounds and Ounces	102
Grams and Kilograms	106
Cups, Pints, Quarts, and Gallons	110
Milliliters and Liters	113
Mixed Review Chapter 8	115
Review Chapter 8	117

Chapter 9: Division

Introduction	119
Division as Making Groups	122
Division and Multiplication	126
Division and Multiplication Facts	130
Dividing Evenly into Groups	133
Division Word Problems	137
Zero in Division	140
When Division Is Not Exact	143
More Practice with the Remainder	146
Mixed Review Chapter 9	148
Review Chapter 9	150

Chapter 10: Fractions

Introduction	152
Understanding Fractions	155
Fractions on a Number Line	159
Mixed Numbers	163
Equivalent Fractions	167
Comparing Fractions 1	170
Comparing Fractions 2	173
Mixed Review Chapter 10	175
Fractions Review	177

Foreword

Math Mammoth Grade 3-A and Grade 3-B worktexts comprise a complete math curriculum for third grade mathematics studies that meets and exceeds the Common Core standards.

Third grade is a time for learning and mastering two (mostly new) operations: multiplication and division within 100. The student also deepens his understanding of addition and subtraction, and uses those in many different contexts, such as with money, time, and geometry.

The main areas of study in Math Mammoth Grade 3 are:

1. Students develop an understanding of multiplication and division of whole numbers through problems involving equal-sized groups, arrays, and area models. They learn the relationship between multiplication and division, and solve many word problems involving multiplication and division (chapters 2, 3, and 9).

2. Students develop an understanding of fractions, beginning with unit fractions. They use fractions along with visual fraction models and on a number line. They also compare fractions by using visual fraction models and strategies based on noticing equal numerators or denominators (chapter 10).

3. Students learn the concepts of area and perimeter. They relate area to multiplication and to addition, recognize perimeter as a linear measure (in contrast with area), and solve problems involving area and perimeter (chapter 7).

4. Students fluently add and subtract within 1,000, both mentally and in columns (with regrouping). They learn to add and subtract 4-digit numbers, and use addition and subtraction in problem solving (chapters 1 and 6).

Additional topics we study are time (chapter 4), money (chapter 5), measuring (chapter 8), and bar graphs and picture graphs (in various chapters).

This book, 3-B, covers place value and 4-digit numbers (chapter 6), geometry (chapter 7), measuring (chapter 8), division (chapter 9), and fractions (chapter 10). The rest of the topics are covered in the 3-A student worktext.

When you use these two books as your only or main mathematics curriculum, they are like a "framework," but you still have a lot of liberty in planning your child's studies. While multiplication and division chapters are best studied in the order they are presented, feel free to go through the geometry, clock, measuring, and fraction sections in a different order. For geometry chapter, the child should already know the multiplication tables.

This might even be advisable if your child is "stuck" on some concept, or is getting bored. Sometimes the brain "mulls it over" in the background, and the concept he/she was stuck on can become clear after a break.

Math Mammoth aims to concentrate on a few major topics at a time, and study them in depth. This is totally opposite to the continually spiraling step-by-step curricula, in which each lesson typically is about a different topic from the previous or next lesson, and includes a lot of review problems from past topics.

This does not mean that your child would not need occasional review. However, when each major topic is presented in its own chapter, this gives you more freedom to plan the course of study *and* choose the review times yourself. In fact, I totally encourage you to plan your mathematics school year as a set of certain topics, instead of a certain book or certain pages from a book.

For review, the download version includes an html page called Make_extra_worksheets_grade3.htm that you can use to make additional worksheets for computation or for number charts. You can also simply reprint some already studied pages.

I wish you success in teaching math!

Maria Miller, the author

Chapter 6: Place Value with Thousands
Introduction

This chapter covers 4-digit numbers (numbers with thousands), and adding and subtracting them. We also study rounding and estimating, which are very important skills for everyday life.

First, students learn place value—breaking 4-digit numbers into their parts (thousands, hundreds, tens, and ones), and comparing. Next, they practice some mental addition and subtraction with 4-digit numbers. The lesson stresses the similarities between adding and subtracting 4-digit numbers and adding and subtracting smaller numbers. This helps build number sense. We also study regrouping in addition and subtraction.

The last major topics in this chapter are rounding numbers to the nearest hundred and estimating. Students also do some problem solving in one lesson.

The Lessons

	page	span
Thousands	8	*4 pages*
Four-Digit Numbers and Place Value	12	*4 pages*
Which Number is Greater?	16	*2 pages*
Mental Adding and Subtracting	18	*4 pages*
Add 4-Digit Numbers with Regrouping	22	*2 pages*
Subtract 4-Digit Numbers with Regrouping	24	*4 pages*
Rounding to the Nearest Hundred	28	*3 pages*
Estimating	31	*3 pages*
Word Problems	34	*3 pages*
Mixed Review Chapter 6	37	*2 pages*
Review Chapter 6	39	*2 pages*

Helpful Resources on the Internet

Base Ten Blocks
Interactive base ten blocks for illustrating numbers up to 10,000. You can also solve problems.
http://www.hoodamath.com/mobile/games/basetenblocks.html

Cookie Dough
Practice either spelling big numbers, or writing the numbers from the words.
http://www.funbrain.com/cgi-bin/nw.cgi?A1=s&A2=10000&A3=1&A12=0
http://www.funbrain.com/cgi-bin/nw.cgi?A1=s&A2=10000&A3=1&A12=1

Sea Life Place Value - Expanded Form
Practice adding numbers in expanded form with this fun, interactive game. With each correct answer, you get to add another beautiful plant or animal to the sea floor!
http://www.free-training-tutorial.com/place-value/sealife/sl-expanded-form.html

Crossword Puzzle – Place Value
http://www.free-training-tutorial.com/word-games/crossword-puzzles-place-value-4.html

Soccer Math Estimating
Practice estimating with four-digit numbers. After each set of problems, you get to try to launch the soccer ball into the net. Choose level 3 and addition or subtraction.
http://www.abcya.com/estimating.htm

Place Value Puzzler
Place value or rounding game. Choose "easy" place value or "easy" rounding for this level.
https://www.funbrain.com/games/place-value

Balloon Pop Math – Order Numbers
Pop the balloons in order from the smallest number to the largest. Choose the number range 1-10,000.
http://www.sheppardsoftware.com/mathgames/placevalue/BPOrder1000.htm

Caterpillar Slider
First, set the max number as high as possible. Then, place the "number leaves" in order on the branch.
http://www.ictgames.com/caterpillar_slider.html

ADDITION AND SUBTRACTION

Place Value Splat
Click on the amounts of hundreds, tens, and ones that equal the given number. Think of regrouping!
http://www.sheppardsoftware.com/mathgames/placevalue/PlaceValuesShapesShoot.htm

Drag and Drop Math – choose subtraction
Practice 4-digit addition or subtraction in columns in this customizable activity.
http://mrnussbaum.com/drag-and-drop-math/

Mental Addition and Subtraction Quiz
http://www.thatquiz.org/tq-1/?-j14g03-l1i-p0

Four-Digit Addition and Subtraction Quizzes
http://www.thatquiz.org/tq-1/?-jg41-l34-p0

http://www.thatquiz.org/tq-1/?-jg42-l34-p0

ROUNDING AND ESTIMATING

Rounding Sharks
Click on the shark that has the number rounded correctly.
http://www.free-training-tutorial.com/rounding/sharks.html

Interactive Rounding Crossword
http://www.free-training-tutorial.com/word-games/crossword-puzzles-rounding-100.html

Maximum Capacity - Estimation
Drag as many gorillas as you can into the elevator without exceeding the weight capacity of the elevator.
http://www.mrnussbaum.com/maximumcapacity.htm

Estimation Game
Estimate the answers by clicking on the number line. Choose "Add 100s" or "Subtract 100s".
http://www.mathsisfun.com/numbers/estimation-game.php

Thousands

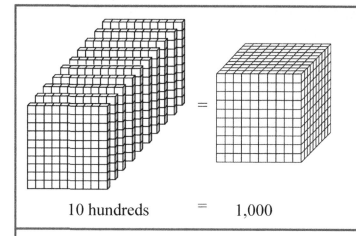

When we take ten hundred-flats and stack them end-to-end, we get *one thousand.*

Ten hundreds = One thousand.

We write a *thousand* as 1000 or 1,000. The comma , is used to separate the "1" of the thousands from the three other digits. It just makes it easier to read.

10 hundreds = 1,000

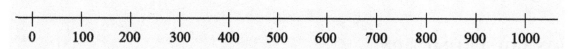

On this number line, you see only whole hundreds marked. In between each two marks are 99 numbers. Imagine those 99 little lines between 300 and 400!

After nine hundred, the next whole hundred is "ten hundreds" or A THOUSAND, 1,000. Remember: *Ten hundreds make a thousand.*

Numbers with four digits are very easy to read. The first of the four digits is in the thousands place. Just read it as "one thousand", "two thousand", "five thousand", and so on.

The rest of the three digits you can read just like you are used to reading three-digit numbers.

One thousand four hundred fifty-nine				Two thousand eighteen				Four thousand seven hundred six			
thou-sands	hund-reds	tens	ones	thou-sands	hund-reds	tens	ones	thou-sands	hund-reds	tens	ones
1	4	5	9	2	0	1	8	4	7	0	6

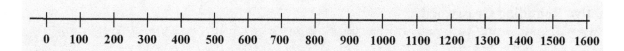

The whole hundreds after one thousand continue as: *one thousand, one thousand one hundred, one thousand two hundred*, etc. Many times, people also read these numbers this way: *a thousand, eleven hundred, twelve hundred, thirteen hundred*, etc.

1. Write the numbers that are illustrated by the models. Sometimes you will need a zero or zeros.

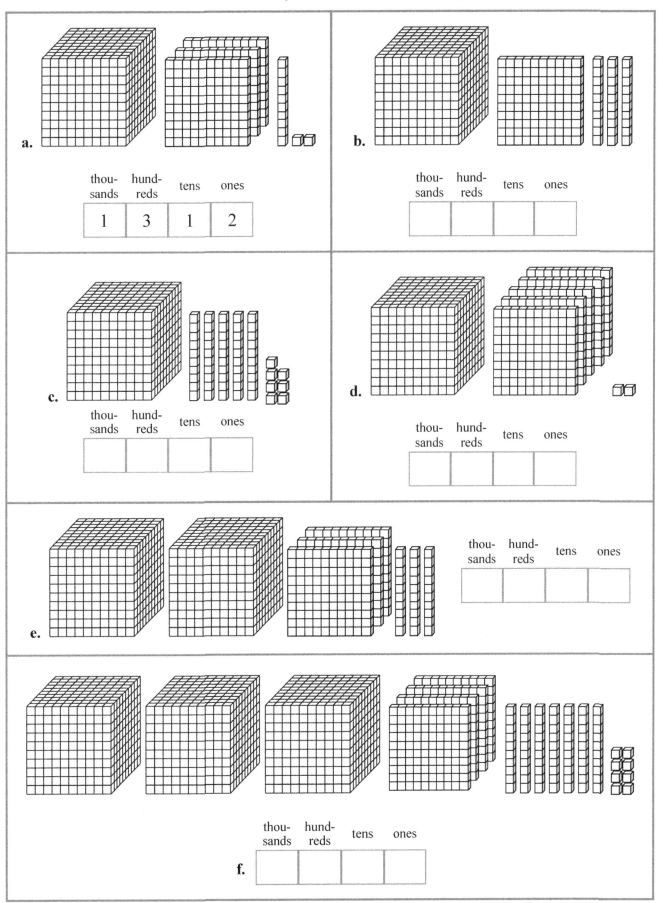

2. Fill in the table.

a. One thousand two hundred fifty-six	b. Three thousand five hundred ninety-four	c. Four thousand six hundred seventeen
thousands: 1, hundreds: 2, tens: 5, ones: 6	thousands: , hundreds: , tens: , ones:	thousands: , hundreds: , tens: , ones:
d. Nine thousand eight hundred twenty-two	**e.** Six thousand two hundred eleven	**f.** Five thousand seven hundred ninety-nine
thousands: , hundreds: , tens: , ones:	thousands: , hundreds: , tens: , ones:	thousands: , hundreds: , tens: , ones:

3. Fill in the table. Now you will need to use a zero or zeros, so be careful!

a. One thousand one	b. Two thousand five	c. Four thousand sixty-one
thousands: 1, hundreds: 0, tens: 0, ones: 1	thousands: , hundreds: , tens: , ones:	thousands: , hundreds: , tens: , ones:
d. Three thousand twelve	**e.** Six thousand two hundred	**f.** Five thousand ninety
thousands: , hundreds: , tens: , ones:	thousands: , hundreds: , tens: , ones:	thousands: , hundreds: , tens: , ones:
g. One thousand one hundred three	**h.** Seven thousand five hundred six	**i.** Five thousand eight hundred
thousands: , hundreds: , tens: , ones:	thousands: , hundreds: , tens: , ones:	thousands: , hundreds: , tens: , ones:
j. Two thousand eleven	**k.** Two thousand three hundred twenty	**l.** Nine thousand thirty-two
thousands: , hundreds: , tens: , ones:	thousands: , hundreds: , tens: , ones:	thousands: , hundreds: , tens: , ones:

4. Fill in the numbers for these number lines.

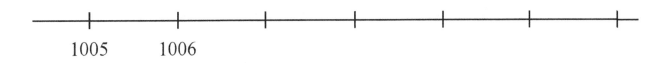

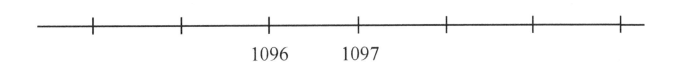

5. Fill in the number chart and count by whole tens.

1 0 1 0	1 0 2 0			
1 0 6 0	1 0 7 0			

Four-Digit Numbers and Place Value

Here the numbers 2467, 1090, and 5602 are written as a *sum* of their different place values.

It is like writing each part of the number out in full: the thousands, the hundreds, the tens, and the ones. **Notice the zeros!** When there are *no* hundreds, or tens, or ones, we write a zero.

thousands	hundreds	tens	ones
2	4	6	7

2000 + 400 + 60 + 7

thousands	hundreds	tens	ones
1	0	9	0

1000 + 0 + 90 + 0

thousands	hundreds	tens	ones
5	6	0	2

5000 + 600 + 0 + 2

1. Fill in the blanks, and write the numbers <u>as a sum</u> of the different place values.

a. 1,034 = ____ thousand ____ hundreds ____ tens ____ ones

= 1000 + ___0___ + __3 0__ + __4__

b. 5,670 = ____ thousand ____ hundreds ____ tens ____ ones

= 5000 + _____ + _____ + _____

c. 3,508 = ____ thousand ____ hundreds ____ tens ____ ones

= _____ + _____ + _____ + _____

d. 8,389 = ____ thousand ____ hundreds ____ tens ____ ones

= _____ + _____ + _____ + _____

e. 9,007 = ____ thousand ____ hundreds ____ tens ____ ones

= _____ + _____ + _____ + _____

f. 7,214 = ____ thousand ____ hundreds ____ tens ____ ones

= _____ + _____ + _____ + _____

2. Fill in the table.

a. Five thousand nine hundred ninety	b. Six thousand sixteen	c. Six thousand three hundred three
T H T O	T H T O	T H T O

d. Eight thousand seven hundred	e. Nine thousand two hundred forty-five	f. Ten thousand
T H T O	T H T O	ten thou-sands T H T O 1 0 0 0 0

3. These numbers are written as sums. Write them in the normal way.

a. 2000 + 90 = _____	b. 8000 + 5 = _____
3000 + 200 = _____	1000 + 80 + 7 = _____
c. 8000 + 200 + 20 = _____	d. 4000 + 50 = _____
2000 + 500 + 90 + 8 = _____	2000 + 800 + 7 = _____

4. What part of these numbers is missing?

a. 5000 + 80 + _____ = 5,083	b. 7000 + _____ + 5 = 7,605
c. _____ + 3000 = 3,050	d. _____ + 700 + 1 = 2,701

5. Write the numbers immediately after and before the given number.

a. _____ 6,049 _____   b. _____ 2,324 _____

c. _____ 1,800 _____   d. _____ 8,809 _____

e. _____ 7,385 _____   f. _____ 9,244 _____

6. These numbers are written as sums, but in a scrambled order! Write them as normal numbers.

a. 4000 + 900 + 7 = _____	**b.** 80 + 500 + 8000 + 6 = _____
c. 2 thousand 7 ones 4 tens	**d.** 2 tens 6 hundred 4 thousand
e. 7 thousand 8 hundred 8 ones	**f.** 5 thousand 6 tens
g. 3 thousand 4 ones	**h.** 5 hundred 9 thousand

7. What part of these numbers is missing?

a. 900 + 2 + _____ = 8,902	**b.** 5000 + 40 + _____ = 5,046
c. _____ + 6000 + 40 = 6,540	**d.** _____ + 4000 + 300 = 4,340

8. Here is a number line from 2,390 to 2,500 with tick-marks for every 10.

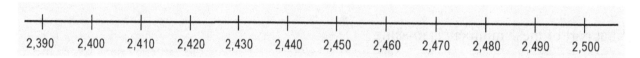

Mark these numbers on the number line (approximately):
2415   2398   2441   2476   2483   2499

9. Draw a number line from 7,650 to 7,750 with tick marks at every 10.

Mark these numbers on the number line (approximately):
7659   7672   7745   7717   7688

14

10. Connect each number inside the puzzle to its whole thousands, hundreds, tens, and ones that it contains. For example, 6,593 is connected to 6,000 and to 500 (for starters). Add the unused numbers from the border to form the missing number inside.

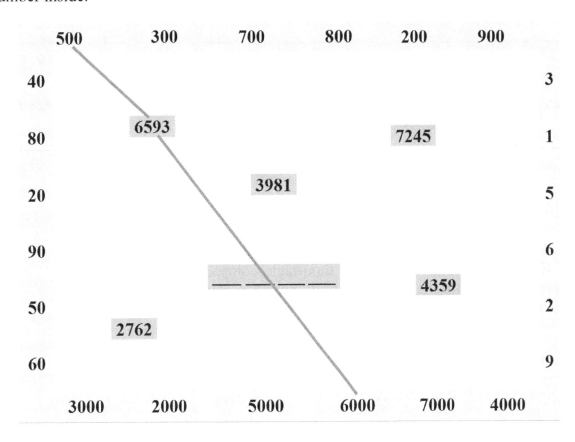

11. Solve the puzzle. Think of breaking the numbers into thousands, hundreds, tens, and ones.

	+		+		+		=	5206
+		+		+		+		
	+		+		+		=	3078
+		+		+		+		
	+		+		+		=	1925
+		+		+		+		
	+		+		+		=	432

= 5022   = 3235   = 1408   = 976

Which Number is Greater?

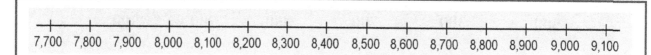

Which is more, 8,011 or 7,987? Place those numbers (approximately) on the number line.

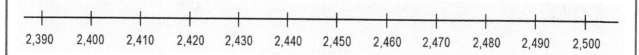

Which is more, 2,395 or 2,402? Place those numbers (approximately) on the number line.

Compare first how many THOUSANDS the numbers have.
Then if they have the same amount of thousands, compare the _____.
Then if they also have the same amount of hundreds, compare the _____.
And if everything else is the same, then compare the ONES.

1. Circle the greatest number.

a. 7,500 6,500 8,500	b. 5,400 5,200 5,700	c. 2,400 4,400 8,400	d. 3,500 3,200 3,300
e. 5,078 5,098 5,100	f. 2,770 2,750 2,760	g. 3,805 3,811 3,809	h. 5,743 5,734 5,721
i. 2,399 4,989 7,011	j. 4,500 6,101 3,099	k. 9,056 9,834 9,275	l. 6,309 9,603 3,609

2. Write < or > between the numbers.

a.	b.	c.	d.
1,050 < 5,095	220 1,020	1,307 1,032	4,012 4,284
2,400 2,750	8,060 6,999	4,906 6,029	5,008 5,040
6,005 4,500	1,007 1,705	5,077 5,570	1,890 1,897

3. One of the three numbers fits on the empty line so that the comparisons are true. Which one? Circle the number (or write it on the line).

a. 6,550 7,601 7,550	b. 2,435 2,338 2,350
7,500 < _____ < 7,600	2,335 < _____ < 2,345
c. 7,099 7,110 7,080	d. 1,232 1,212 1,223
7,089 < _____ < 7,100	1,203 < _____ < 1,222
e. 8,752 8,502 7,802	f. 4,216 4,111 4,096
8,459 < _____ < 8,510	4,097 < _____ < 4,200
g. 1,809 1,908 1,890	h. 3,489 3,589 3,458
1,806 < _____ < 1,812	3,469 < _____ < 3,579

4. Compare. Write < , > , or = in the box.

a.   700 + 50 ☐ 700 + 30 + 4   b.   500 + 6000 ☐ 6000 + 500

c.   20 + 3000 ☐ 300 + 2000   d.   900 + 8 ☐ 9000 + 8

e.   4000 + 80 ☐ 80 + 4 + 800   f.   30 + 6000 + 3 ☐ 300 + 60 + 3000

g.   800 + 7000 + 2 ☐ 700 + 80 + 7000

h.   500 + 3000 + 80 + 6 ☐ 6 + 80 + 500 + 3000

5. Write the numbers in order from smallest to greatest. The number line can help.

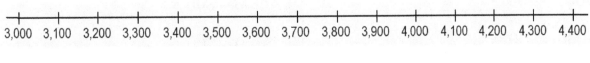

3899   3040   4330   4203   4003

_____ < _____ < _____ < _____ < _____

Mental Adding and Subtracting

1. Skip-count by one hundred.

   5000, 5100, _____, _____, _____, _____

   2800, 2900, _____, _____, _____, _____

   > Solve the problem 2700 + 200 by adding 27 + 2 = 29.
   > That is because 2700 is 27 hundreds, and 200 is 2 hundreds.
   > So, the answer is 29 hundreds, or 2900 (tag two zeros to 29).

2. Whole hundreds! It means they end in two zeros.

a. ten hundreds = 1000	**b.** _____ hundreds = 5600
21 hundreds = _____	_____ hundreds = 7900
42 hundreds = _____	_____ hundreds = 8000

3. Add and subtract. Think of the whole hundreds.

a. 5000 + 200 = _____	**b.** 2900 + 200 = _____
5100 + 400 = _____	3100 + 300 = _____
c. 6800 + 400 = _____	**d.** 5600 − 200 = _____
3800 + 800 = _____	4500 − 300 = _____

   ```
   +----+----+----+----+----+----+----+----+----+----+----+----+----+----+
   7,700 7,800 7,900 8,000 8,100 8,200 8,300 8,400 8,500 8,600 8,700 8,800 8,900 9,000 9,100
   ```

e. 8000 − 200 = _____	**f.** 7900 − 800 = _____
8000 − 700 = _____	8500 − 700 = _____
g. 2200 − 600 = _____	**h.** 9800 − 700 = _____
3500 − 600 = _____	1300 − 300 = _____

4. Complete the next whole thousand.

a. 600 + ____ = 1 0 0 0	b. 6600 + ____ = ____
2500 + ____ = 3 0 0 0	2400 + ____ = ____
c. 500 + ____ = ____	d. 8200 + ____ = ____
9200 + ____ = ____	7300 + ____ = ____

5. Add and subtract. Again, think of the whole hundreds.

a. 5000 + 1200 = ____	b. 2700 + 3200 = ____
5100 + 2400 = ____	3100 + 6300 = ____
c. 2500 + 2500 = ____	d. 1600 + 1700 = ____
3500 + 3500 = ____	3600 + 4500 = ____

6. Mark flew in an airplane 3,200 km from his hometown to Phoenix.
Then he drove in a car 300 km to see his grandma.
How long was his trip one way?

How long was his trip both ways?

7. Solve (find the number that the symbol stands for).

a. 1,200 + △ = 1,500 △ = ____	b. 5,600 + △ = 6,100 △ = ____	c. 7,700 − △ = 7,200 △ = ____
d. 9,000 − △ = 2,500 △ = ____	e. △ − 400 = 6,500 △ = ____	f. △ − 4,400 = 3,000 △ = ____

Unknown in addition or subtraction
We can write ? or ☐ or some other symbol for the **unknown thing** in an addition or subtraction sentence. Study the examples.
A used van costs $4,000. Dad pays $1,700 of it now and the rest later. How much is left to pay later? Maybe you can solve this in your head, but even so, let's learn to write a number sentence with an unknown. We can write an addition: $1,700 + ? = $4,000. We could also write a subtraction: $4,000 − $1,700 = ? Solution: He will pay $2,300 later.

8. Write an addition or a subtraction for each problem. Use ? or ☐ for the unknown thing.

a. An expensive camera costs $5,000. Ashley has saved $3,700. How much more money does she still need?
b. A jogging track is 4,200 feet long. Through it, there is a shortcut that shortens it to only 3,100 feet. How much does the shortcut shorten the track?
c. Josh jogs around the track using the shortcut, three times. How many feet did he jog in total?
d. From his paycheck, Denny pays $500 in taxes. Then he pays $700 as rent. Now he has $1,000 left. How much is his paycheck?
e. A car dealer was going to sell a car for $800, but then he doubled the price. Then a customer came, and he told the customer, "I will take some money off the price." So, the customer paid $1,200. How much did the dealer take off the price?

9. Count by tens.

   a. 4000, 4010, _____, _____, _____, _____

   b. _____, _____, 1740, 1750, _____, _____

   c. _____, _____, 3370, 3380, _____, _____

10. Add and subtract. Compare the problems.

a. 100 + 20 = _____	b. 220 + 40 = _____
5100 + 20 = _____	4220 + 40 = _____
c. 140 − 90 = _____	d. 230 − 30 = _____
4140 − 90 = _____	4230 − 30 = _____

11. Add and subtract. Below, you can write a helping problem without the thousands.

| a. 4,980 + 20 = _____   980 + 20 = _____ | b. 7,210 + 90 = _____ |
| c. 7,760 − 30 = _____ | d. 5,540 + 50 = _____ |

Puzzle Corner

What numbers can go into the puzzle?

4,550	−		+		= 4,560
−		+		−	
	+		+		= 50
+		−		+	
	+		+		= 100
= 4,580		= 30		= 60	

21

Add 4-Digit Numbers with Regrouping

Add thousands in their own column. Regrouping (carrying) is done the same way as before. You might have to regroup <u>three</u> times: in the tens, in the hundreds, and in the thousands.

```
  1 1 1
  5 8 7 9           3 3 7 1           4 7 6 8
+ 2 5 4 4         + 3 9 9 8         + 2 6 5 5
─────────         ─────────         ─────────
  8 4 2 3
```

Here we regroup three times.

Finish these examples yourself and ask your teacher to check.

1. Add. It helps to add those numbers first which make ten (if any)!

   a.   5 0 9 1   **b.**   2 3 9 3   **c.**   5 8 0 2
      +   5 1 0   + 4 7 1 6   + 1 8 7 0

   d.   6 0 9 8   **e.**   2 2 5 5   **f.**   3 6 2
         1 0 3 4   3 4 5   2 3 8 9
      +   2 5 4   + 2 1 7 0   + 4 0 6 7

   g.   4 5 6   **h.**   1 6 5 9   **i.**   3 7 3
         7 3 2 8   1 9 9   2 8 8
         1 1 3 4   2 6 7   5 2 1 7
      +   5 5 4   + 6 0 3 7   + 3 3 9 9

2. Add. Be careful to line up the ones, tens, hundreds, and thousands.

a. 34 + 2,382 + 391 + 77 + 3,409

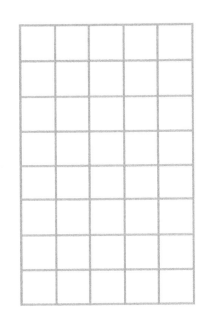

b. 450 + 349 + 3,822 + 39 + 8

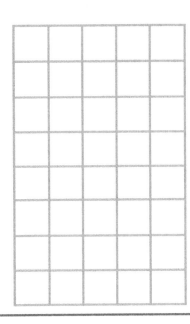

3. Solve.

a. Find the total bill when David buys two adult airplane tickets for $1,250 each and three children's tickets for $698 each.

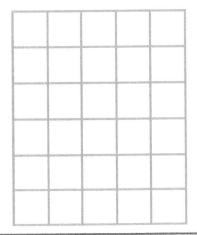

b. An airplane ticket for an adult is $655. A child's ticket is $200 cheaper. Find the cost of two adult and two children's tickets.

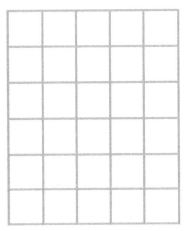

Puzzle Corner

What numbers are missing?

```
  3 □ 5 □            2 9 □ □
+ □ 1 □ 9          + □ □ 3 6
---------          ---------
  9 0 8 1            5 4 1 7
```

Subtract 4-Digit Numbers with Regrouping

You cannot subtract 4 tens from 3 tens. So, regroup one hundred as 10 tens.	You cannot subtract 2 hundreds from 0 hundreds, so regroup 1 thousand as 10 hundreds. Complete.	Check by adding.
$$\begin{array}{r}5139\\-2244\\\hline 5\end{array} \quad \begin{array}{r}0\;13\\5\cancel{1}\cancel{3}9\\-2244\\\hline 5\end{array}$$	$$\begin{array}{r}0\;13\\5\cancel{1}\cancel{3}9\\-2244\\\hline 95\end{array} \quad \begin{array}{r}10\\4\;\cancel{0}\;13\\\cancel{5}\cancel{1}\cancel{3}9\\-2244\\\hline 95\end{array}$$	$$\begin{array}{r}\\+2244\\\hline\end{array}$$

1. Subtract. Check by adding.

a. $\quad 5091$
$\quad\; -\;510 \quad +$ _____

b. $\quad 2913$
$\quad -1716 \quad +$ _____

c. $\quad 8402$
$\quad -1378 \quad +$ _____

d. $\quad 6881$
$\quad\; -\;911 \quad +$ _____

e. $\quad 6546$
$\quad -3490 \quad +$ _____

f. $\quad 9080$
$\quad -5025 \quad +$ _____

g. $\quad 4509$
$\quad -1116 \quad +$ _____

h. $\quad 6209$
$\quad -2065 \quad +$ _____

Regrouping with zeros

You cannot subtract 5 ones from 4, so you need to regroup.	There are no tens nor hundreds, so regroup 1 thousand as 10 hundreds.	Then regroup 1 hundred as 10 tens.	Lastly, regroup 1 ten as 10 ones. There are already 4 ones, so you get 14 ones. Subtract.
9 0 0 4 − 3 6 5 5	8 10 9̶ 0̶ 0 4 − 3 6 5 5	9 8 1̶0̶ 10 9̶ 0̶ 0̶ 4 − 3 6 5 5	9 9 8 1̶0̶ 1̶0̶ 14 9̶ 0̶ 0̶ 4̶ − 3 6 5 5

2. Subtract. Check by adding.

a. 4 0 0 2
   − 2 2 1 6   + _____

b. 6 1 2 0
   − 3 8 4 4   + _____

c. 4 3 0 3
   − 4 0 0 8   + _____

d. 7 0 1 1
   −   9 1 2   + _____

e. 5 0 1 3
   − 2 4 9 0   + _____

f. 9 0 0 1
   − 4 0 7 5   + _____

g. 3 3 0 0
   − 1 4 0 1   + _____

h. 8 0 0 5
   − 1 7 7 9   + _____

3. Solve.

a. 4,908 − 203 − 1,420

b. 9,000 − (3,450 + 593)

c. 3,924 + 291 + 2,932 − 2,910

d. Solve what number the triangle represents:

△ − 5,480 = 1,027

4. Three villages form a triangle. The distance between Riverville and Middleville is 3,200 meters, between Middleville and Highville 1,900 m, and between Highville and Riverville 4,200 m.

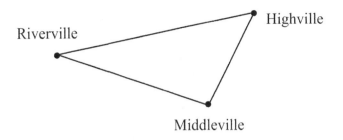

   a. Write the distances between the villages on the map.

   b. What is the total distance from Riverville to Highville to Middleville and back to Riverville?

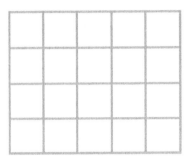

   c. How many more meters is a round trip from Middleville to Riverville and back, than from Highville to Middleville and back?

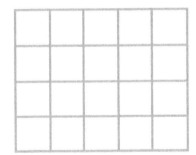 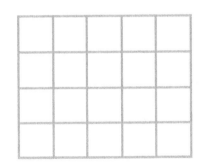 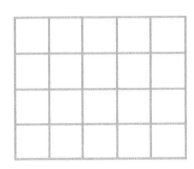

Puzzle Corner   What numbers are missing?

```
  3 □ 5 □          8 9 □ □          6 □ □ 9          □ □ 0 3
- □ 1 □ 9        - □ □ 3 6        - □ 2 2 □        - 2 8 □ □
---------        ---------        ---------        ---------
  1 6 6 4          4 7 1 8          2 7 8 4          4 1 3 6
```

Rounding to the Nearest Hundred

When we round **to the nearest hundred**, the numbers "residing" in the red areas (up to 850) on the number line are rounded to 800. The numbers in the blue areas are rounded to 900.

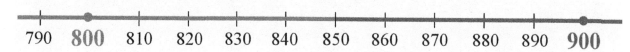

Again, distance matters. Numbers from 801 to 849 are closer to 800 than to 900. Numbers from 851 to 899 are closer to 900 than to 800. And the "middle guy," 850, is rounded up to 900: 850 ≈ 900.

1. Mark the numbers as dots on the number line (approximately) and round them to either 800 or 900.

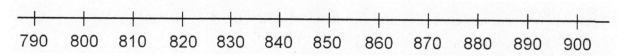

a.	b.	c.	d.
804 ≈ _____	825 ≈ _____	860 ≈ _____	808 ≈ _____
859 ≈ _____	850 ≈ _____	847 ≈ _____	892 ≈ _____

When rounding to the **nearest hundred**, look at the TENS DIGIT of the number.
- If that digit is 0, 1, 2, 3, or 4, you will *round down* to the previous hundred.
- If that digit is 5, 6, 7, 8, or 9, you will *round up* to the next hundred.

7**2**8 ≈ ? The tens digit is 2, so round down: 728 ≈ 700	4**7**1 ≈ ? The tens digit is 7, so round up: 471 ≈ 500	9**5**6 ≈ ? The tens digit is 5, so round up: 956 ≈ 1000

2. Round these numbers to the nearest hundred.

a.	b.	c.	d.
416 ≈ _____	529 ≈ _____	670 ≈ _____	254 ≈ _____
837 ≈ _____	960 ≈ _____	557 ≈ _____	147 ≈ _____

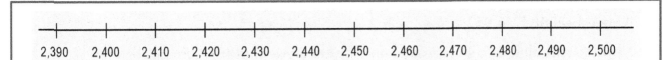

Things work similarly on this number line when rounding <u>to the nearest hundred</u>:

- Numbers up to 2,450 (the middle number) are rounded down to 2,400.
- Numbers after 2,450 are rounded up to 2,500.
- And the "middle guy," 2,450, is rounded up.

3. Round these numbers to 2,400 or to 2,500. You can use the number line above to help you.

| a. 2,412 ≈ _____ | b. 2,429 ≈ _____ | c. 2,478 ≈ _____ | d. 2,490 ≈ _____ |

Rounding rules

When rounding to the **nearest hundred**, look at the TENS DIGIT of the number.

If the tens digit is 0, 1, 2, 3, or 4, you will *round down* to the previous hundred.	If that digit is 5, 6, 7, 8, or 9, you will *round up* to the next hundred.
728 ≈ ? The tens digit is 2, so round down: 728 ≈ 700	471 ≈ ? The tens digit is 7, so round up: 471 ≈ 500
Notice: the hundreds digit **does not change**.	Notice: the hundreds digit **increases by one**, from 4 to 5.

Whether you round up or down, the tens and ones digits *change to zeros*.

4. Round these numbers to the nearest hundred.

a. 6,216 ≈ _____	b. 5,923 ≈ _____	c. 1,670 ≈ _____
d. 8,254 ≈ _____	e. 8,019 ≈ _____	f. 2,157 ≈ _____
g. 1,772 ≈ _____	h. 6,849 ≈ _____	i. 801 ≈ _____
j. 255 ≈ _____	k. 9,562 ≈ _____	l. 3,501 ≈ _____

> **Note especially!** $4{,}9\underline{5}2 \approx ?$
>
> Now the tens digit is 5, so we round up. The hundreds digit (9) increases by one (to ten hundreds). But we can't change 9 to 10 or we would get 41052!
>
> Instead, those ten hundreds make a *new thousand*, so the thousands digit is incremented by one. It is as if '49' changes to '50'. So, we get $\underline{\mathbf{4{,}9}}52 \approx \underline{\mathbf{5{,}0}}00$

5. Round these numbers to the nearest hundred.

a. 6,986 ≈ _____	b. 5,973 ≈ _____	c. 1,981 ≈ _____
d. 4,945 ≈ _____	e. 2,932 ≈ _____	f. 9,969 ≈ _____
g. 966 ≈ _____	h. 9,982 ≈ _____	i. 798 ≈ _____

6. Round these numbers to the nearest hundred. Place each answer in the cross-number puzzle.

   Across:
   a. 2,264 ≈ _____
   b. 4,973 ≈ _____
   c. 4,248 ≈ _____
   d.   545 ≈ _____

   Down:
   e. 3,709 ≈ _____
   f.   672 ≈ _____
   g. 5,370 ≈ _____
   h. 8,816 ≈ _____

7. Fill in, using rounded numbers.

   a. Usually, Mary receives about _____ spam emails daily, but on 5/9 she got about _____ spams.

   b. During the work week from 5/7 till 5/11 she received about _____ spams.

Spam Emails Mary Received		
Date	Spams	round to nearest 100
Mo 5/7	125	
Tu 5/8	97	
Wd 5/9	316	
Th 5/10	118	
Fr 5/11	106	

Estimating

| Estimating means that we don't calculate the exact answer, but instead we use *rounded numbers* in the calculation. The answer we get that way is called the **estimate**. It is close to the real answer. | Estimation:

303 + 2,278
↓ ↓
300 + 2,300 = 2,600 | Exact calculation:

$\,^{1}$
$3\;0\;3$
$+\,2\;2\;7\;8$
$\overline{2\;5\;8\;1}$ |

1. Estimate the results of these additions and subtractions by rounding the numbers to the nearest hundred. On the right, calculate the exact answers.

a. Estimation: 569 + 234 ↓ ↓ *600* + ___ = ___	Exact calculation: $5\;6\;9$ $+\,2\;3\;4$ $\overline{}$
b. Estimation: 8,155 + 424 ↓ ↓ ___ + ___ = ___	Exact calculation: $8\;1\;5\;5$ $+\,4\;2\;4$ $\overline{}$
c. Estimation: 577 − 125 ↓ ↓ ___ − ___ = ___	Exact calculation: $5\;7\;7$ $-\,1\;2\;5$ $\overline{}$
d. Estimation: 7,028 − 465 ↓ ↓ ___ − ___ = ___	Exact calculation: $7\;0\;2\;8$ $-\,4\;6\;5$ $\overline{}$

2. Estimate these additions and subtractions by rounding the numbers to the nearest hundred. On the right, calculate the exact answer.

a. Estimation:

5,171 + 568
↓ ↓

____ + ____ = _____

Exact calculation:

```
  5 1 7 1
+   5 6 8
---------
```

b. Estimation:

4,162 + 3,439
↓ ↓

____ + ____ = _____

Exact calculation:

```
  4 1 6 2
+ 3 4 3 9
---------
```

c. Estimation:

7,577 − 2,947
↓ ↓

____ − ____ = _____

Exact calculation:

```
  7 5 7 7
− 2 9 4 7
---------
```

d. Estimation:

756 + 4,178 + 836
↓ ↓ ↓

____ + ____ + ____ = _____

Exact calculation:

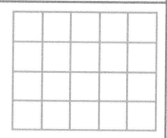

e. Estimation: 8,295 − 5,538 − 1,150 − 924
↓ ↓ ↓ ↓

____ − ____ − ____ − ____ = _____

e. Exact calculation:

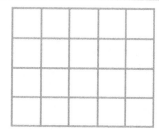 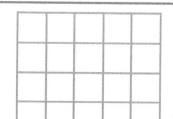 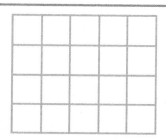

Now check. Were your estimations close to the real answers?

3. Elisa's estimates are kind of far from her answers. Find where Elisa has made an error.

a. Elisa's estimation: $6{,}540 - 259$ $\downarrow \quad\quad \downarrow$ $6500 - 300 = 6200$	Elisa's work: $\quad\ 6\ 5\ 4\ 0$ $-\quad\ 2\ 5\ 9$ $\overline{\quad\ 6\ 3\ 9\ 1}$
b. Elisa's estimation: $3{,}834 - 1{,}260$ $\downarrow \quad\quad \downarrow$ $3800 - 1300 = 2500$	Elisa's work: $\quad\ 3\ 8\ 3\ 4$ $-\ 1\ 2\ 6\ 0$ $\overline{\quad\ 2\ 9\ 7\ 4}$
c. Elisa's estimation: $3{,}874 + 1{,}990$ $\downarrow \quad\quad \downarrow$ $3900 + 2000 = 5900$	Elisa's work: $\quad\ 3\ 8\ 7\ 4$ $+\ 1\ 9\ 9\ 0$ $\overline{\quad\ 4\ 8\ 6\ 4}$

4. Here is a typical Colombian shopping list. The amounts are in *pesos*.

   Estimate the total of this shopping list. First round each price to the nearest hundred, and then add.

   The exact bill is 8823 pesos. Compare your estimate and the exact total. Was your estimation close?

		Rounded numbers:
rice 1 kg	2750	
parsley	449	
potatoes	1876	
tomatoes	1564	
bananas	1238	
onions	946	
TOTAL	8823	

Word Problems

Example. Find the change when Daniel buys a lawn mower for $1,589 and pays with $2,000. Also, estimate the answer using rounded numbers.

To estimate the answer, we round 1,589 to 1,600. The estimation is $2,000 − $1,600 = $400. So, Daniel's change should be about $400.

We use the estimated answer ($400) to check if our final answer is reasonable. Subtracting the exact numbers (on the right), we get $411 as the change. That is reasonable because it is close to our estimate of $400.

```
        9  9
    1 10 10 10
    2  0  0  0
  − 1  5  8  9
       4  1  1
```

1. Latoya bought a fridge for $1,158 and a freezer for $745. She paid with $2,000. What was her change?

   Also, estimate the answer:

   _____

   _____

2. A new motorcycle costs $8,740 and a used one $1,295. What is the price difference?

   Also, estimate the answer using rounded numbers:

   _____

   _____

3. Using the digits 1, 2, 3, and 4, build the largest and the smallest number possible.

   What is the difference of the two?

4. Can you buy three air conditioners at $979 each, with $3,000?
   If yes, how much will be left over?
   If not, how much more money would you need?

5. A store owner bought four washer/dryer machines for $1,109 each. Then he got $500 off of his total bill (a discount). Find what he had to pay.

   Also, estimate the answer:

   _____

   _____

6. Jack is a fisherman. The pictograph shows how many kilograms of fish he caught last week.

   Each 🐟 represents 200 kg of fish.

Fish Caught	
Monday	🐟🐟🐟
Wednesday	🐟🐟🐟🐟🐟🐟▸
Friday	🐟🐟🐟🐟🐟▸
Sunday	🐟🐟🐟▸

   a. How many kilograms of fish did he catch on Wednesday?

   b. How many kilograms of fish did he catch on Friday?

   c. How many more kilograms of fish did he catch on Friday than on Monday?

   d. How many kilograms of fish did he catch in total during this week?

7. Alex checked the price of a certain TV in four different stores.

   a. Draw a bar graph from his results.

   b. How much is the difference between the most and the least expensive TV?

	Price
Bob's TV Store	$525
The Nerdy Store	$564
Home Express	$632
Lion Appliances	$599

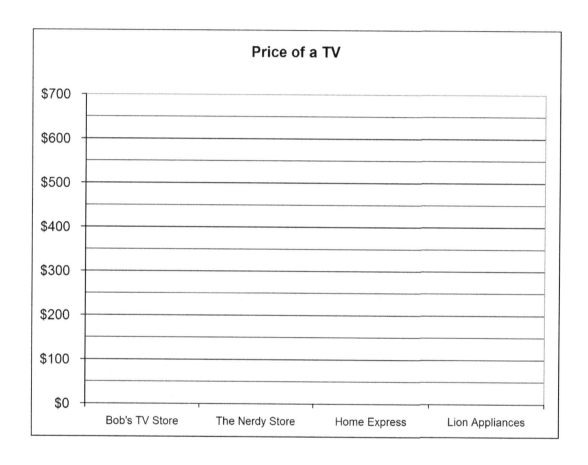

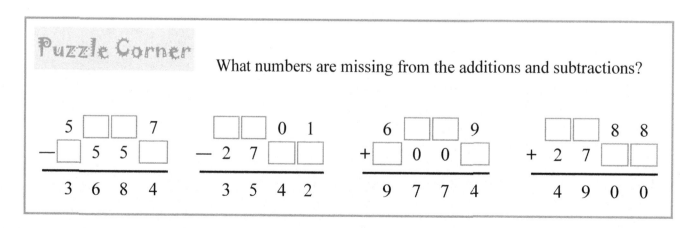

Puzzle Corner — What numbers are missing from the additions and subtractions?

Mixed Review Chapter 6

1. Subtract and compare the problems.

a. 56 – 7 = _____	b. 72 – 9 = _____	c. 83 – 8 = _____
256 – 7 = _____	672 – 9 = _____	283 – 8 = _____

2. Write a number sentence for these word problems, and solve them.

a. How many legs do six chickens and four dogs have in total?
_____

b. Nine classrooms in the school have four windows each, and one has only three. How many windows are there in total?
_____

3. Round the numbers to the nearest ten.

a. 574 ≈ _____	b. 895 ≈ _____	c. 604 ≈ _____	d. 56 ≈ _____
342 ≈ _____	255 ≈ _____	427 ≈ _____	998 ≈ _____

4. Write the Roman numerals using normal numbers.

a. III	b. XIX	c. LXXXV	d. XLIII
VIII	XXIV	LIII	CXXV
XIV	LX	XL	CCLXXI

5. Write using Roman numerals.

a. 15	b. 21	c. 56	d. 90
19	43	65	99

6. Find the missing factors.

a. ___ × 6 = 24	b. 7 × ___ = 49	c. 8 × ___ = 64	d. ___ × 9 = 36
___ × 6 = 54	7 × ___ = 35	8 × ___ = 48	___ × 9 = 72
___ × 6 = 42	7 × ___ = 56	8 × ___ = 32	___ × 9 = 45

7. How much time passes?

a. From 5:00 AM to 10:20 AM	b. From 8:30 AM to 1:00 PM
c. From 7 PM to 6 AM	d. From 10:55 AM to 3:55 PM

8. Find the total cost of buying the items listed. Line up the numbers carefully for adding.

| $2.90 | $1.45 | $7.50 | $18.49 | $6.32 | $1.50 |

a. scissors and crayons	b. pencils, a calculator, and a book	c. two books and two calculators

Review Chapter 6

1. Fill in the table.

a. Seven thousand two hundred forty	**b.** Six thousand five	**c.** Two thousand twenty-nine
T H T O	T H T O	T H T O

2. These numbers are written as sums. Write them in the normal way.

a. 7000 + 500 + 3 = _____

3000 + 90 = _____

b. 30 + 1000 + 7 = _____

400 + 6000 = _____

3. Compare. Write <, >, or = in the box.

a. 7000 + 50 ☐ 5000 + 7

b. 500 + 4 + 6000 ☐ 6000 + 400 + 5

c. 80 + 3000 ☐ 3000 + 200

d. 400 + 80 ☐ 8000 + 40

4. Add and subtract mentally.

a. 1,200 + 700 = _____

400 + 6,800 = _____

b. 3,600 − 300 = _____

4,200 − 500 = _____

c. 7,200 + _____ = 8,000

8,000 − _____ = 7,100

d. 3,400 + 1,500 = _____

7,500 + 800 = _____

5. Solve (find the number that the symbol stands for).

a. 3,400 + △ = 4,100

△ = _____

b. △ − 600 = 9,200

△ = _____

c. 10,000 − △ = 8,500

△ = _____

6. Round these numbers to the nearest hundred.

a.	b.	c.	d.
872 ≈ _____	5,253 ≈ _____	6,034 ≈ _____	2,739 ≈ _____

7. Add and subtract. Estimate first by rounding the numbers to the nearest hundred.

a. Estimate:

2,540 + 1,803
↓   ↓

_____ + _____ = _____

Calculate exactly: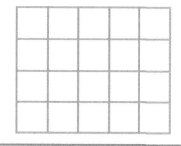

b. Estimate:

6,581 − 736
↓   ↓

_____ − _____ = _____

Calculate exactly: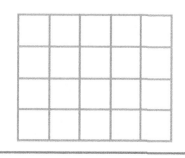

8. Solve the word problems.

a. Dad bought wood for construction for $1,616, paint for $278, and other materials for $969. Find his total bill.

Also, estimate the answer using rounded numbers.

My estimate: _____

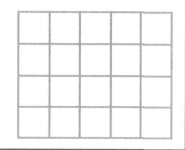

b. You have $5,000 to spend. First, you buy a pump for $278 and then some cement for $1,250. How much do you have left after that?

Also, estimate the answer using rounded numbers.

My estimate: _____

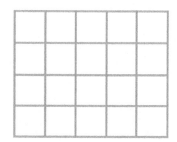 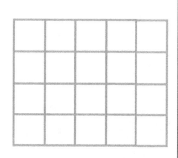

Chapter 7: Geometry
Introduction

The seventh chapter of *Math Mammoth Grade 3* deals with geometry. The emphasis is on two new concepts: area and perimeter.

First, we study and review shapes in one lesson where the student divides shapes into new ones, and also encounters some tilings (a.k.a. tessellations). Next, we study in more detail about some quadrilaterals, namely squares, rectangles, and rhombi (plural of rhombus).

Then comes the focus of this chapter: perimeter and area. Students find perimeters of polygons, including finding the perimeter when the side lengths are given, and finding an unknown side length when the perimeter is given.

They learn about area, and how to measure it in square inches, square feet, square centimeters, square meters, or just square units if no unit of length is specified.

Students also relate area to the operations of multiplication and addition. They learn to find the area of a rectangle by multiplying the side lengths, and to find the area of rectilinear figures by dividing them into rectangles and adding the areas.

We also study the distributive property "in disguise." This means using an area model to represent $a \times (b + c)$ as being equal to $a \times b$ plus $a \times c$. The expression $a \times (b + c)$ is the area of a rectangle with side lengths a and $(b + c)$, which is equal to the areas of two rectangles, one with sides a and b, and the other with sides a and c.

Multiplying by Whole Tens is a lesson about multiplication such as 3×40 or 90×7. It is put here so that students can use their multiplication skills to calculate areas of bigger rectangles.

Then we solve many area and perimeter problems. That is necessary so that students learn to distinguish between these two concepts. They also get to see rectangles with the same perimeter and different areas or with the same area and different perimeters.

Lastly, we touch on solids, such as cubes, rectangular prisms, pyramids, cones, and cylinders, and study their faces, edges, and vertices. You can make paper models for them from the printouts provided in the curriculum. Alternatively, you can buy them, usually made in plastic. Search on the Internet for "geometric solids."

The Lessons

	page	span
Shapes	46	*4 pages*
Some Special Quadrilaterals	50	*3 pages*
Perimeter	53	*3 pages*
Problems with Perimeter	56	*3 pages*
Getting Started with Area	59	*2 pages*
More About Area	60	*4 pages*
Multiplying by Whole Tens	65	*4 pages*
Area Units and Problems	69	*4 pages*
Area and Perimeter Problems	73	*2 pages*
More Area and Perimeter Problems	75	*3 pages*
Solids	78	*2 pages*
Mixed Review Chapter 7	80	*2 pages*
Geometry Review	82	*2 pages*

Helpful Resources on the Internet

Use these online resources as you see fit to supplement the main text.

SHAPES

Shapes Splat
Get points by clicking on the correct shapes.
http://www.sheppardsoftware.com/mathgames/earlymath/shapes_shoot.htm

Shapes Identification Quiz from ThatQuiz.org
An online quiz in a multiple-choice format, asking to identify common two-dimensional shapes. You can modify the quiz parameters to your liking.
www.thatquiz.org/tq-f/math/shapes/

Quadrilateral Shapes Shoot
Practice identifying quadrilaterals. You can choose relaxed or fast mode.
http://www.sheppardsoftware.com/mathgames/geometry/shapeshoot/QuadShapesShoot.htm

Matching Shapes
Pair all the tiles by matching the polygons with their proper names.
http://www.mathplayground.com/matching_shapes.html

Interactive Quadrilaterals
Drag the corners to play with squares, rectangles, rhombi, and more.
http://www.mathsisfun.com/geometry/quadrilaterals-interactive.html

Polygon Playground
Drag various colorful polygons to the work area to make your own creations!
http://www.mathcats.com/explore/polygons.html

Shape Cutter
Draw any shape (polygon), cut it, and manipulate the cut pieces. You can have the computer mix them up, and then try to recreate the original shape.
http://illuminations.nctm.org/ActivityDetail.aspx?ID=72

Patch Tool
An online activity where the student designs a pattern using geometric shapes.
http://illuminations.nctm.org/ActivityDetail.aspx?ID=27

Tangram Puzzles for Kids
Use the seven pieces of the Tangram to form the given puzzle. Complete the puzzle by moving and rotating the seven shapes.
http://www.abcya.com/tangrams.htm

Tangram Game
Arrange the five geometrical shapes that are given to form various shapes.
http://www.tangramgames.co.uk/tangramgameA/

Interactivate! Tessellate
An online, interactive tool for creating your own tessellations. Choose a shape, then edit its corners or edges. The program automatically changes the shape so that it will tessellate (tile) the plane. Then push the tessellate button to see your creation!
http://www.shodor.org/interactivate/activities/Tessellate

Online Kaleidoscope
Create your own kaleidoscope creation with this interactive tool.
https://web.archive.org/web/20160309222840/http://www.zefrank.com/dtoy_vs_byokal/

AREA AND PERIMETER

Free Worksheets for Area and Perimeter
Create customizable worksheets for the area and the perimeter of rectangles. Options include using images, generating word problems, or problems where the student writes an expression for the area using the distributive property.
http://www.homeschoolmath.net/worksheets/area_perimeter_rectangles.php

FunBrain: Shape Surveyor Geometry Game
An easy game that practices finding either the perimeter or area of rectangles.
http://www.funbrain.com/poly/index.html

Perimeter Shapes Shoot Game
"Shoot" the shapes that have the given perimeter.
http://www.sheppardsoftware.com/mathgames/geometry/shapeshoot/PerimeterShapesShoot.htm

Perimeter at Gordons
Work out the perimeter of the shapes. There are many options to choose from.
http://www.wldps.com/gordons/Perimeter.swf

Shape Explorer
Find the perimeter and area of odd shapes on a rectangular grid.
http://www.shodor.org/interactivate/activities/ShapeExplorer/

Area of Rectangle
Drag the corners of the rectangle and see how the side lengths and areas change.
http://illuminations.nctm.org/ActivityDetail.aspx?ID=46

Build a Robot
Collect six parts to build your own robot by answering questions about perimeter.
http://www.learnalberta.ca/content/me3us/flash/lessonLauncher.html?lesson=lessons/12/m3_12_00_x.swf

Area Shapes Shoot Game
Click on the shapes that show the given area.
http://www.sheppardsoftware.com/mathgames/geometry/shapeshoot/AreaShapesShoot.htm

Math Playground: Party Designer
You need to design areas for the party, such as a crafts table, food table, seesaw, and so on, so they have the given perimeters and areas.
https://www.mathplayground.com/PartyDesigner/index.html

Zoo Designer
You have been hired to design five enclosures for the animals at a local zoo. Use your knowledge of area and perimeter to design the correct enclosures and to earn your ZooDesigner Points.
http://mrnussbaum.com/zoo/

Area Blocks
Cover your grid with shapes before your opponent does.
http://www.mathplayground.com/area_blocks.html

Area and Perimeter Builder
Create your own rectangular shapes using colorful blocks and explore the relationship between perimeter and area. You can choose to show the side lengths to understand how a perimeter works. You can also use two work areas (grids) to compare the area and perimeter of two shapes side-by-side. Lastly, challenge yourself in the game screen to build shapes or find the area of various figures.
http://phet.colorado.edu/sims/html/area-builder/latest/area-builder_en.html

Math Playground: Measuring the Area and Perimeter of Rectangles
Amy and her brother, Ben, explain how to find the area and perimeter of rectangles and show you how changing the perimeter of a rectangle affects its area. After the lesson, you will use an interactive ruler to measure the length and width of 10 rectangles, and to calculate the perimeter and area of each.
http://www.mathplayground.com/area_perimeter.html

XP Math: Find Perimeters of Parallelograms
This online quiz shows you parallelograms and rectangles, and you need to calculate the perimeter, including typing in the right unit, and not using the altitude of the parallelogram.
http://www.xpmath.com/forums/arcade.php?do=play&gameid=10

Area and the Distributive Property Quiz
Use area models to represent the distributive property in finding area of rectangles.
https://www.khanacademy.org/math/cc-third-grade-math/cc-third-grade-measurement/cc-third-grade-area-distributive-property/e/area-and-the-distributive-property

MULTIPLY BY MULTIPLES OF TEN

Multiplying by Multiples of Ten
Drag the correct answer over to its problem.
http://mrnussbaum.com/grade_3_standardsmultbytens/

Multiplication Quiz
Practice your skills of multiplying by multiples of ten in this 10-question online quiz.
http://www.thatquiz.org/tq-1/?-jkg04-lc-p0

SOLIDS

Identify solids
Select the name and drop it on the correct solid.
http://www.softschools.com/math/geometry/shapes/solids/games/

Identify solids
Click to identify the partially buried 3-dimensional shapes.
http://www.primaryinteractive.co.uk/online/longshape3d.html

Geometric Solids
Manipulate various geometric solids. Color the solid to investigate properties such as the number of faces, edges, and vertices.
http://illuminations.nctm.org/ActivityDetail.aspx?ID=70

Under the Sea
First, choose 3-D shapes. Then, click on a "magic crystal" to start an activity.
http://www.learnalberta.ca/content/me3usa/flash/index.html?goLesson=14

2-D and 3-D Shapes
Learn about different solids and see them rotate.
http://coolsciencelab.com/2D_3D_shapes.swf

Shapes

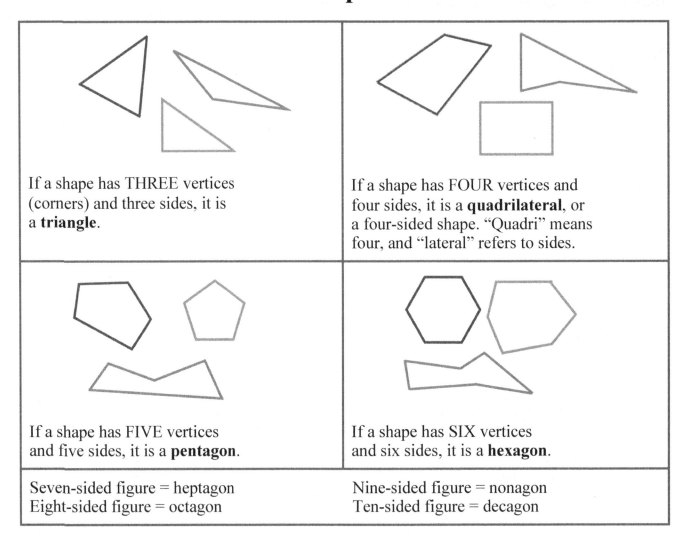

If a shape has THREE vertices (corners) and three sides, it is a **triangle**.

If a shape has FOUR vertices and four sides, it is a **quadrilateral**, or a four-sided shape. "Quadri" means four, and "lateral" refers to sides.

If a shape has FIVE vertices and five sides, it is a **pentagon**.

If a shape has SIX vertices and six sides, it is a **hexagon**.

Seven-sided figure = heptagon
Eight-sided figure = octagon

Nine-sided figure = nonagon
Ten-sided figure = decagon

1. Draw two <u>pentagons</u> here by drawing dots and connecting them with lines. Remember, your pentagons do not have to look "regular" or nice. You can draw them to look "funny," too, as long as they have five sides and five vertices.

2. What shape is formed if you place the bolded sides of the two figures together?
   You can trace the shapes and cut them out.

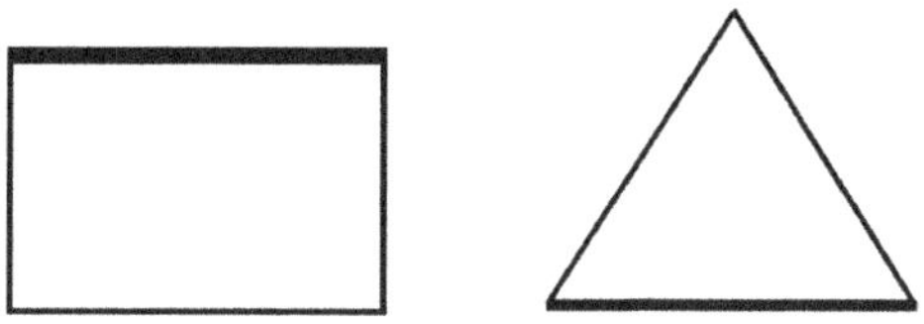

a. _____

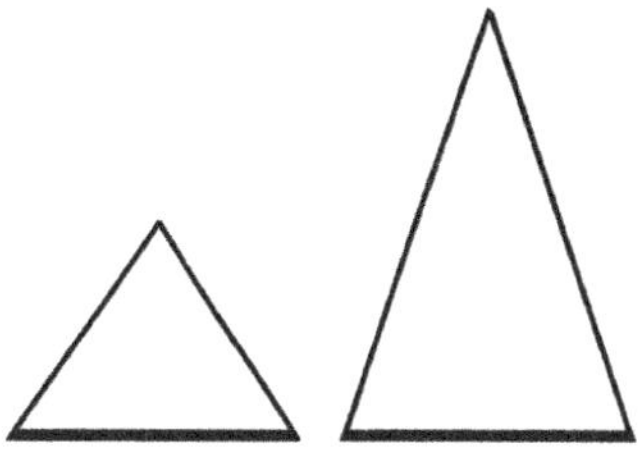

b. _____

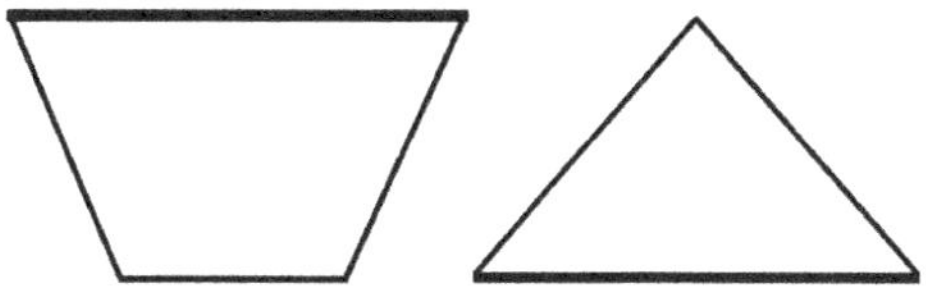

c. _____

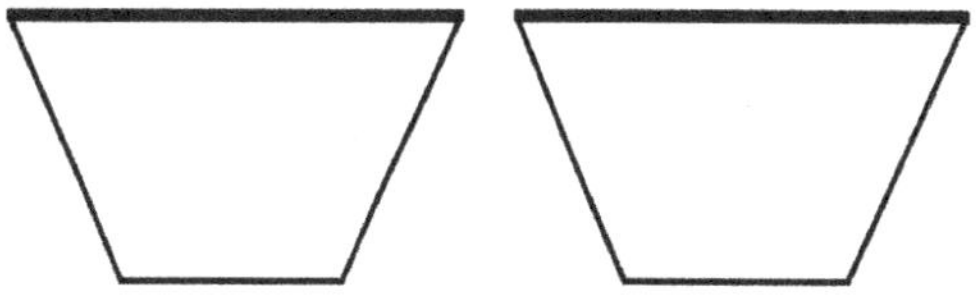

d. _____

3. Draw a straight line or lines through the shape and divide it into other shapes!

a. a square and a rectangle

b. a triangle and a pentagon

c. three rectangles

d. two quadrilaterals that are not rectangles

e. two parts that are exactly the same shape

f. four triangles

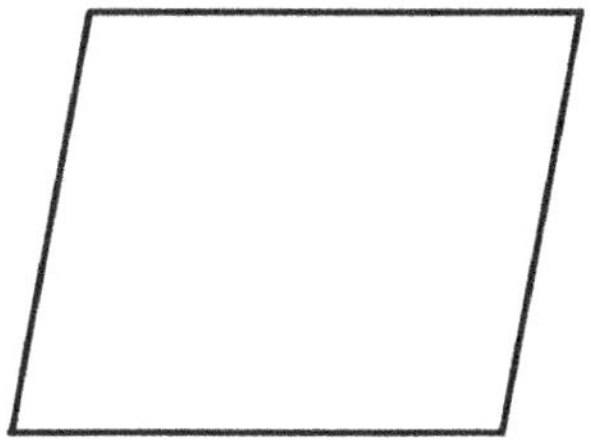

g. four triangles

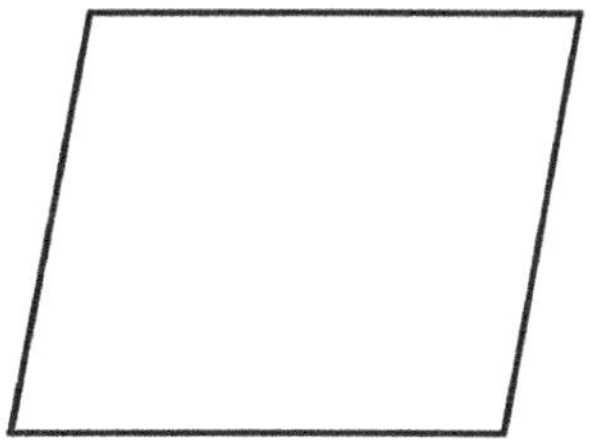

h. a triangle and a pentagon

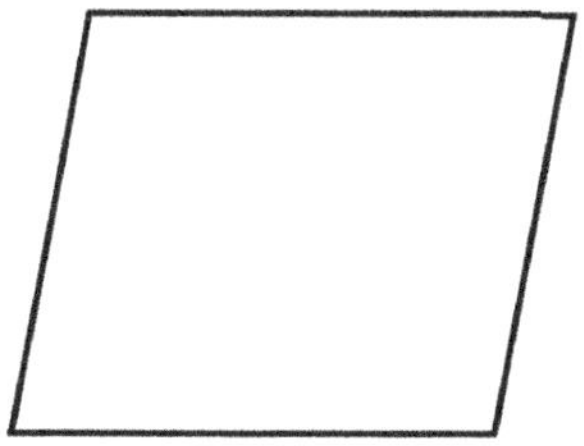

i. four quadrilaterals

4. Divide the pentagon and the hexagon into new shapes using one straight line. Notice: your line does NOT have to go from corner to corner. Write what new shapes you get.

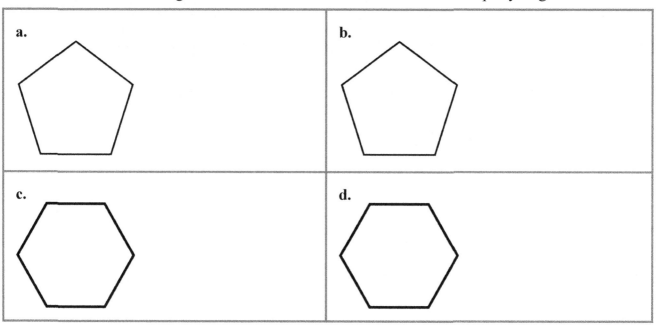

a.

b.

c.

d.

5. Continue the tilings so they fill the grids, and name what shape(s) are used in the tiling.

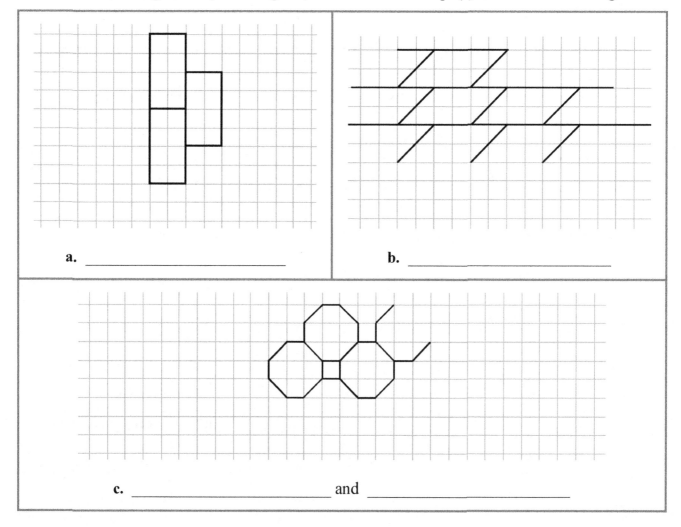

a. _____

b. _____

c. _____ and _____

6. Design your own tilings here.

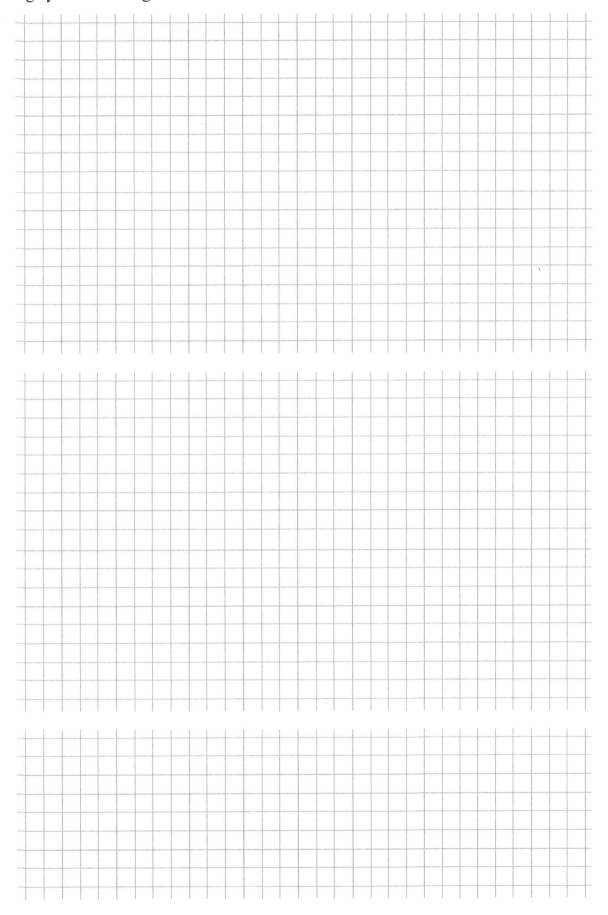

Some Special Quadrilaterals

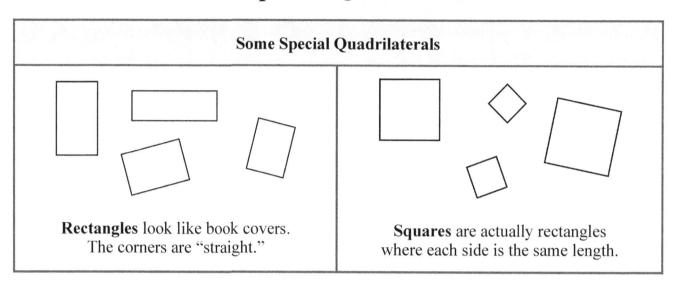

1. Draw three different rectangles and three different squares.

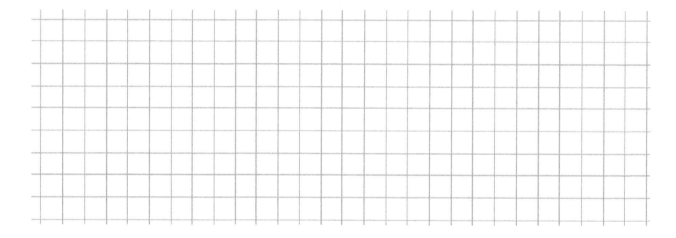

2. Draw three quadrilaterals that are NOT squares nor rectangles.

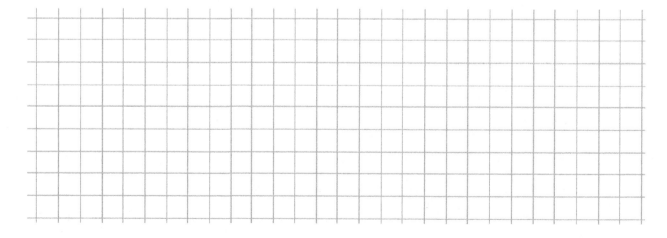

A **rhombus** is a quadrilateral where each of the four sides has the same length. A rhombus is also called a diamond-shape or a diamond in common language.

The corners of a rhombus do not have to be "straight" like the corners of a square, but they can be.

The plural of rhombus is **rhombi**.

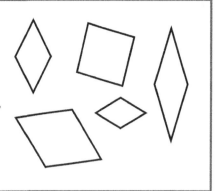

3. You can make a rhombus by taking four popsicle sticks or pencils or other sticks of the same length.

   Arrange the four sticks into a diamond shape. Now, change it slightly to get another rhombus. Make a skinny one, a less skinny one, and so on. You can even make a square!

   You can also play with rhombi on this web page. Choose "rhombus." Just drag the dots!

   http://www.mathsisfun.com/geometry/quadrilaterals-interactive.html

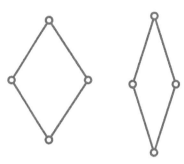

4. A square or a rhombus?

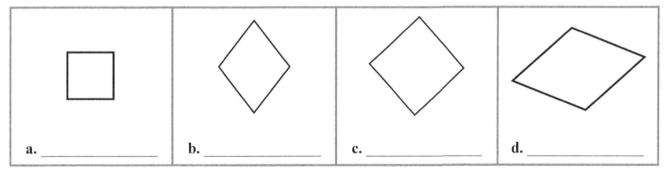

a. _____ b. _____ c. _____ d. _____

5. Is a square *also* a rhombus? Read the definitions again:

   Squares are rectangles (with straight corners) where each side is the same length.

   A **rhombus** is a quadrilateral where each of the four sides has the same length.

So, is a square *also* a rhombus?

Why or why not?

6. Color all the rectangles green, squares blue, rhombi red, and other quadrilaterals yellow. Or, choose your own colors.

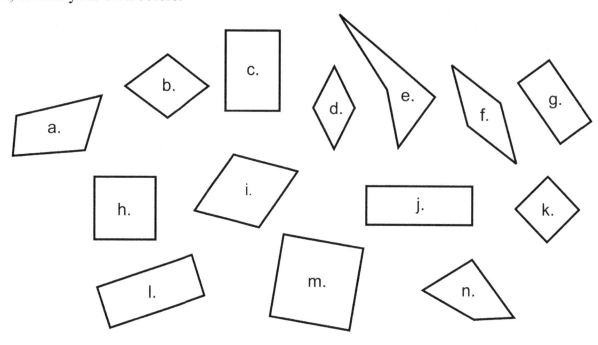

7. This is a tiling with rhombi. Continue it! Use pretty colors.

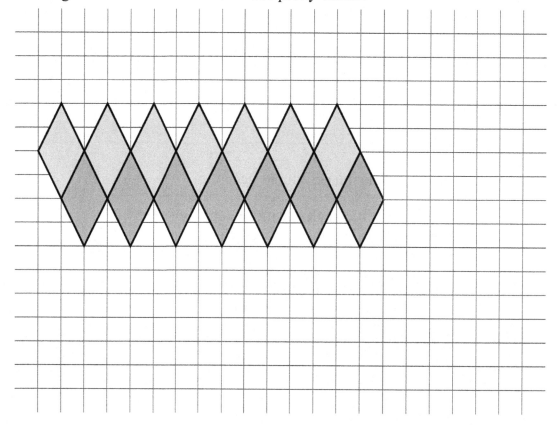

Perimeter

Perimeter means the "walk-around measure," or the distance you go if you walk all the way around the figure.

The word comes from the Greek word *perimetros*. *Peri* means 'around' and *metros* means 'measure'.

To find the perimeter of this rectangle, count the units as you go around the figure. You can think of running or hopping around the figure.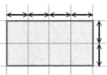

The units are marked with little arrows in the picture. The top side is four units long. The right side is two units long. Make sure you understand that!

So, what is the perimeter? _____ units

Here it is trickier to count those little units. Be careful!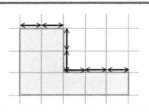

How many units is the perimeter? _____ units

1. Find the perimeter of these figures. Your answer will be so many units. P means perimeter.

| a. 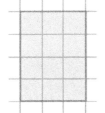 P = _____ *units* | b. 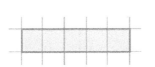 P = _____ | c. 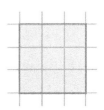 P = _____ |
| d.  P = _____ | e. 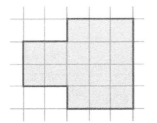 P = _____ | f. 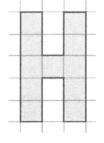 P = _____ |

2. Measure with a ruler to find the perimeter of these figures in centimeters.

a.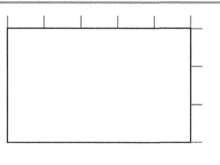

P = _____ cm

b.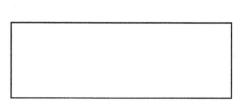

P = _____ cm

c.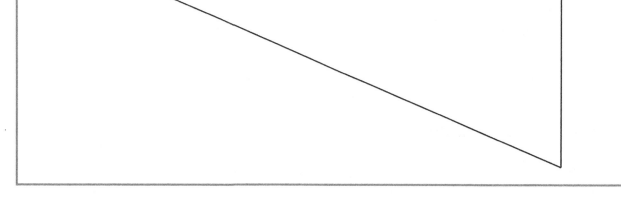

3. Measure with a ruler to find the perimeter of these figures in inches.

a.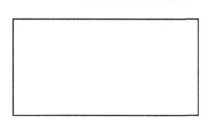

P = _____ in.

b.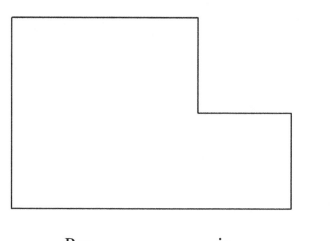

P = _____ in.

You can trace the ruler below and tape it on an existing ruler or cardboard!
Or cut it out after you have finished the neighboring page.

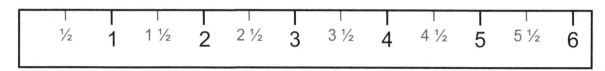

To find the perimeter, simply **add all the side lengths.** How many units is the perimeter of the triangle on the right? It is 8 + 9 + 10 units, or _____ units.	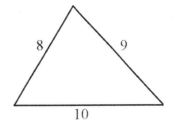
Often you need to figure out some side lengths that are not given. What side lengths are not given? The perimeter is _____ cm.	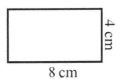

Don't forget the unit of measurement in your answer.

If the side lengths are in centimeters, the perimeter will be so-many *centimeters*.

If the side lengths are "plain numbers" without any particular unit, then the perimeter is so-many *units*.

4. Find the perimeter. Notice: some side lengths are not given! Don't forget to use either "cm," "in." or "units" in your answer.

a.

P = _____ *units*

b.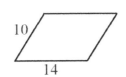

P = _____

c.

P = _____

d.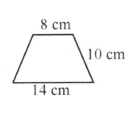

P = _____

e.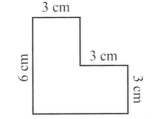

P = _____

f.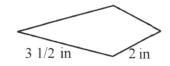

P = _____

5. Find the perimeter....

   a. ...of a square with 7-in. sides

   b. ...of a square with 13-cm sides

Problems with Perimeter

The perimeter of a rectangle is 30 cm. Its one side is 9 cm. How long is the other side?

We can write a "how many more" addition, or an addition with an unknown:

9 + _?_ + 9 + _?_ = 30

You could guess and check to solve it. But, there is also an easier way. Just think: the two sides, 9 and _?_, form **half** of the perimeter. So, 9 + _?_ = 15.

Thinking either way, we can solve that _?_ = 6 cm.

1. Solve. Write an addition with an unknown for each problem.

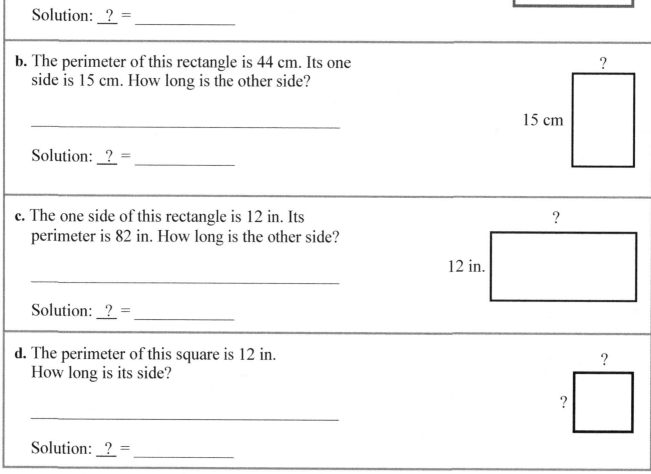

a. The perimeter of this rectangle is 20 cm. Its one side is 6 cm. How long is the other side?

_____

Solution: _?_ = _____

b. The perimeter of this rectangle is 44 cm. Its one side is 15 cm. How long is the other side?

_____

Solution: _?_ = _____

c. The one side of this rectangle is 12 in. Its perimeter is 82 in. How long is the other side?

_____

Solution: _?_ = _____

d. The perimeter of this square is 12 in. How long is its side?

_____

Solution: _?_ = _____

2. Solve.

   a. The perimeter of this square is 44 cm.
   How long is the side of the square?

   b. Find the perimeter of this square with 12-inch sides.

   c. Find the perimeter of this L-shape. Notice that some side lengths are not given.

3. The parking lot of a school is in the shape shown here. Each little square in the image has a side of 10 feet. What is the perimeter of the parking lot?

   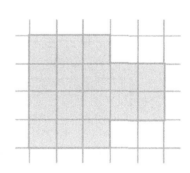

4. Kyle's house measures 25 feet wide and 35 feet long. What is its perimeter?

5. Mandy wants a rectangular garden with a perimeter of 18 meters. One side of the garden is 3 m. How long should the other side be?

6. Draw many different rectangles that all have a perimeter of 24 units. Then, write the side lengths of those rectangles in the table.

   Hint: The two sides of the rectangle form half of the perimeter, which is 12 units.

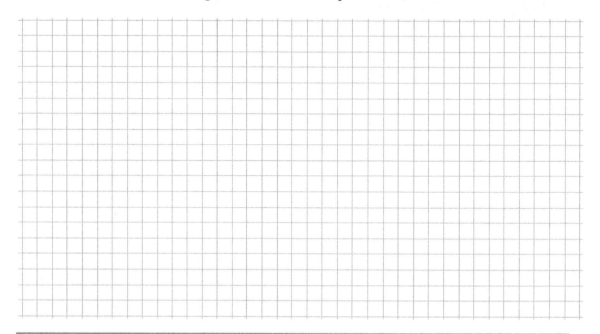

One side	Other side	Perimeter
3 units	9 units	24 units
		24 units
		24 units
		24 units

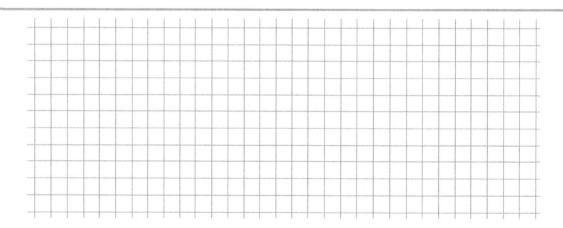

Draw a shape here that is **not** a rectangle, and that has a perimeter of

a. 8 units   b. 10 units   c. 14 units

Puzzle Corner

Getting Started with Area

How many little squares do you need to cover this rectangle?

That is its *area*. Area has to do with covering, and it is measured in little squares, which we call *square units*.

The area of this rectangle is _____ square units.

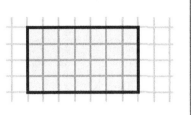

1. How many square units is the area of these figures?

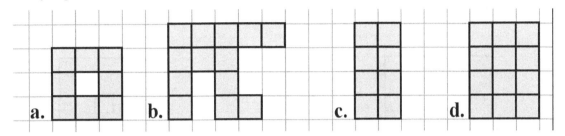

a. The area is _____ square units.

b. The area is _____ square units.

c. The area is _____ square units.

d. The area is _____ square units.

You can use multiplication to find the area of a rectangle. Notice how there are rows and columns of squares!

There are 3 rows, and 8 columns. We multiply 3 × 8 = 24.

The area of this rectangle is 24 square units.

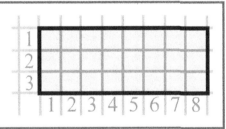

2. Write a multiplication to find the area. "A" means area.

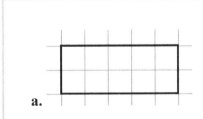

a.

____ × ____ = _____

A = _____ square units.

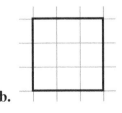

b.

____ × ____ = _____

A = _____ square units.

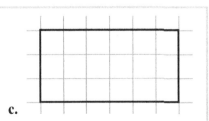

c.

____ × ____ = _____

A = _____ square units.

59

3. Find the areas of these figures.

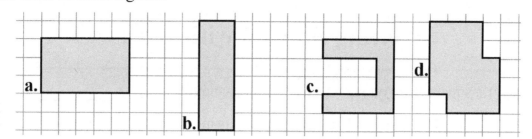

a. The area is _____ square units.

b. The area is _____ square units.

c. The area is _____ square units.

d. The area is _____ square units.

4. Find the areas.

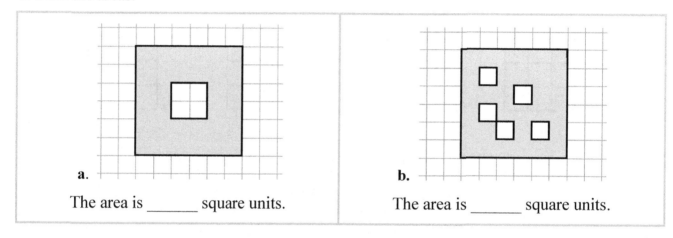

a. The area is _____ square units.

b. The area is _____ square units.

5. Draw two rectangles or squares with an area of 16 square units.

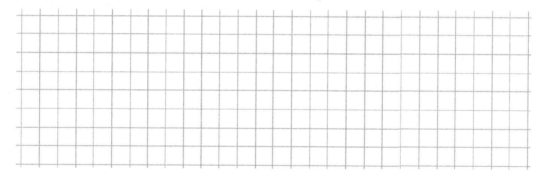

6. Draw two rectangles with an area of 24 square units.

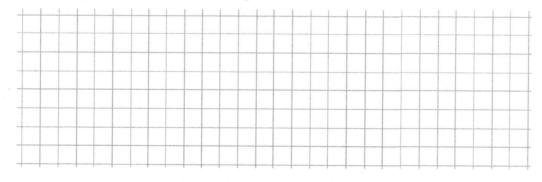

More about Area

To find the area of this figure, we can divide the shape into two rectangles. We then use two multiplications, and add their results.

$3 \times 2 + 3 \times 5 = 6 + 15 = 21$ square units

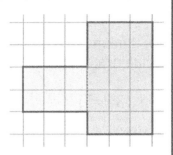

Here, can you think how to use multiplication and *subtraction* to find the shaded area? Don't look at the answer (below) yet! Think first!

It is $4 \times 5 - 2 \times 2 = 20 - 4 = 16$ square units

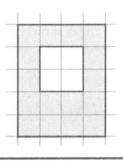

1. Write two multiplications to find the total area.

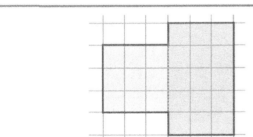

a.

___ × ___ + ___ × ___ = ___

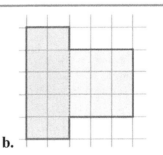

b. _____

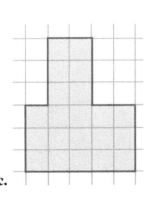

c. _____

d. _____

The total area of this rectangle is 3 × 8 = 24 square units. But notice: we can write the longer side of the rectangle as a sum (3 + 5). Then, its area would be written as 3 × (3 + 5).

But if we think of it as two rectangles, we can write the area as 3 × 3 + 3 × 5.

So, thinking of it as a one rectangle or two rectangles, we get:

3 + 5

3 × (3 + 5) = 3 × 3 + 3 × 5
area of the   area of the   area of the
whole rectangle   first part   second part

2. Write a number sentence for the total area, thinking of one rectangle or two.

a.

___ × (___ + ___) = ___ × ___ + ___ × ___
area of the   area of the   area of the
whole rectangle   first part   second part

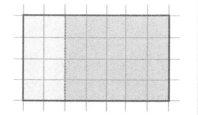

b.

___ × (___ + ___) = ___ × ___ + ___ × ___
area of the   area of the   area of the
whole rectangle   first part   second part

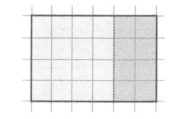

c.

___ × (___ + ___) = ___ × ___ + ___ × ___
area of the   area of the   area of the
whole rectangle   first part   second part

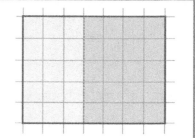

d.

___ × (___ + ___) = ___ × ___ + ___ × ___

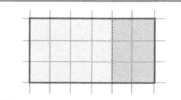

e.

___ × (___ + ___) = ___ × ___ + ___ × ___

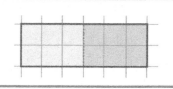

3. Now it's your turn to draw the rectangle. Fill in.

a.

3 × (2 + 4) = ___ × ___ + ___ × ___

area of the whole rectangle   area of the first part   area of the second part

b.

5 × (1 + 4) = ___ × ___ + ___ × ___

area of the whole rectangle   area of the first part   area of the second part

c.

4 × (3 + 1) = ___ × ___ + ___ × ___

area of the whole rectangle   area of the first part   area of the second part

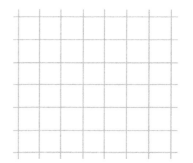

d.

___ × (___ + ___) = 3 × 2 + 3 × 1

area of the whole rectangle   area of the first part   area of the second part

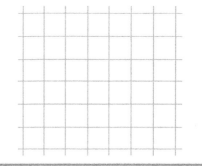

e.

___ × (___ + ___) = 2 × 5 + 2 × 2

area of the whole rectangle   area of the first part   area of the second part

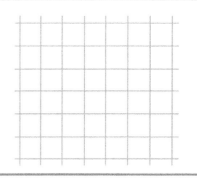

4. Find the areas of the figures.

a. Find the shaded area. Write a number sentence for the area.

_____

_____

b. Find the shaded area.
Think what operations you can use this time.
Write a number sentence for the area.

_____

_____

c. Find the shaded area *(not including the school)*. Write a number sentence for the area.

_____

_____

Puzzle Corner The area of this shape is 32 square units. Your task is to write a number sentence for the area.

Multiplying by Whole Tens

1. Fill in the missing parts of the multiplication table of 10. Think of counting by tens!

9 × 10 = _____	14 × 10 = _____	19 × 10 = _____
10 × 10 = _____	15 × 10 = _____	20 × 10 = _____
11 × 10 = _____	16 × 10 = _____	21 × 10 = _____
12 × 10 = _____	17 × 10 = _____	22 × 10 = _____
13 × 10 = _____	18 × 10 = _____	23 × 10 = _____

There is a pattern: *Every answer ends in* _____. Also, there is something special about the number you multiply times 10, and the answer. Can you see that?

SHORTCUT

To multiply any number by *ten*, write the number, and place a zero after it.

For example: 78 × 10 = 780   or   10 × 49 = 490

---

2. Multiply.

a. 10 × 11 = _____	**b.** 10 × 99 = _____	**c.** 82 × 10 = _____
56 × 10 = _____	18 × 10 = _____	10 × 0 = _____

Note: If the number you multiply by 10 *ends* in zero, you still need to place a zero after it.

For example: 30 × 10 = 300

---

3. Multiply.

a. 10 × 5 = _____	**b.** 10 × 90 = _____	**c.** 17 × 10 = _____
10 × 50 = _____	100 × 9 = _____	17 × 1 = _____

This rectangle illustrates the multiplication 7 × 20.
It has 7 rows and 20 columns.

We could COUNT the little squares to find its area.
Or, we could solve 7 × 20 by adding 20 repeatedly.

But here is yet a different way to think about it:
Let's divide this big rectangle into TWO
smaller rectangles that are the size 7 × 10 each.

Each of the two rectangles has an area of 7 × 10 = 70.
So, in total their area is 70 + 70 = 140.

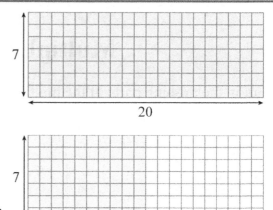

4. Solve.

a. Solve 8 × 20 by dividing this rectangle into TWO equal parts.

Parts: ____ × _____ and ____ × _____. The total area is _____.

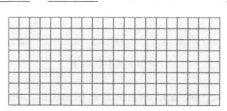

b. Solve 5 × 30 by dividing this rectangle into THREE equal parts.

Parts: ____ × _____ and ____ × _____ and ____ × _____. The total area is _____.

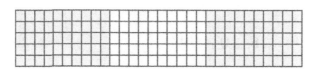

c. Solve 7 × 30 by dividing this rectangle into THREE equal parts.

Parts: ____ × _____ and ____ × _____ and ____ × _____. The total area is _____.

d. Solve 4 × 40 by dividing this rectangle into parts.

Parts: _____. The total area is _____.

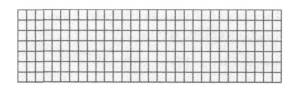

> We can solve multiplication problems, such as 5 × 60, by repeated addition.
>
> 5 × 60 = 60 + 60 + 60 + 60 + 60
>
> (60 added five times)

5. Solve these multiplications by repeated addition. But also look for a pattern and a shortcut. Can you find it?

| a. 3 × 40 = _____ | b. 2 × 80 = _____ | c. 4 × 40 = _____ |
| d. 5 × 30 = _____ | e. 5 × 70 = _____ | f. 3 × 80 = _____ |

> Here's another idea for solving multiplication problems, such as 5 × 60.
>
> Notice: 60 is equal to 6 × 10, isn't it?
> So, to solve 5 × 60, we can multiply 5 × 6 × 10.
>
> And 5 × 6 × 10 is the same as 30 × 10.
> Then, 30 × 10 is just 30 with a zero placed at the end... or 300.

6. Break each multiplication into another where you multiply three numbers, one of them being 10. Multiply and fill in.

a. 7 × 90 = _7_ × _9_ × 10 = _63_ × 10 = _____	b. 4 × 80 = ____ × ____ × 10 = ____ × 10 = _____
c. 6 × 40 = ____ × ____ × 10 = ____ × 10 = _____	d. 9 × 90 = ____ × ____ × 10 = ____ × 10 = _____
e. 30 × 6 = 10 × ____ × ____ = 10 × ____ = _____	f. 80 × 3 = 10 × ____ × ____ = 10 × ____ = _____

Study the **shortcut** for multiplying by whole tens.	
Example 1. 6 × 20 Multiply **6 × 2** = 12. Place a zero after 12, to get 120.	**Example 2.** 90 × 7 Multiply **9 × 7** = 63. Place a zero after 63, to get 630.

7. Multiply using the shortcut.

a. 7 × 70 = _____	b. 6 × 80 = _____	c. 40 × 7 = _____
d. 50 × 4 = _____	e. 70 × 3 = _____	f. 3 × 90 = _____

8. This rectangle is 7 units high and 80 units long. What is its area?

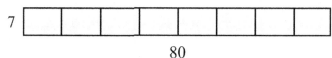

9. This rectangle is divided into 8 equal parts. What is the area of each small part?

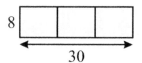

10. Find the total area of this rectangle, and also the area of each little part.

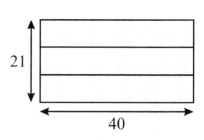

11. Find the total area.

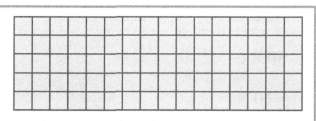

Figure out a way or two ways to solve 5 × 16 *without* counting all the squares.

Puzzle Corner

Area Units and Problems

Area is always measured in *squares of some size*. To find the area of a shape, we check how many squares are needed to cover the shape.

 Each side of this square measures 1 centimeter. It is a special square. It is called **a square centimeter**. We can use it to measure areas of other shapes.

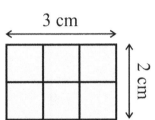 We need 6 square centimeters to cover this rectangle. So, its area is just that: 6 square centimeters. We abbreviate this as 6 cm². The elevated ² indicates the "squaring."

We can also use *multiplication* to find the area:

$$3 \text{ cm} \times 2 \text{ cm} = 6 \text{ cm}^2$$

1. Write a multiplication for the area of each rectangle. Measure the sides of the rectangles in centimeters using a ruler. Don't forget the units (cm and cm²)!

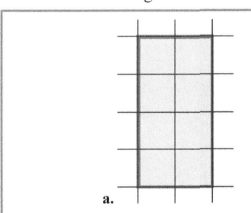

a. A = _____ cm × _____ cm = _____ cm²

b. A = _____ cm × _____ cm = _____ cm²

c. A = _____

d. A = _____

69

Each side of this square measures 1 inch. It is also a special square. It is called **one square inch**, abbreviated as 1 sq. in. or 1 in^2.

We can use it to measure areas of other shapes.

2. Find the area of each rectangle. Measure in inches using a ruler. Don't forget the unit for the area.

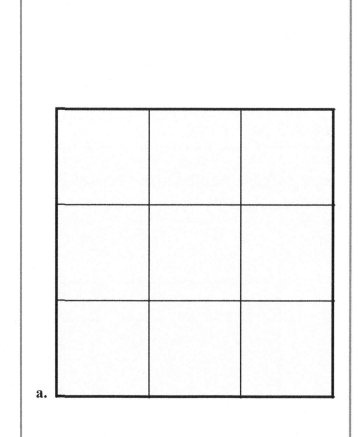

a. A = _____ in. × _____ in. = _____ in^2

b. A = _____ in. × _____ in. = _____ in^2

c. A = _____

The following pictures are *not* to scale. They show some other square units for area.

This is one square foot or 1 ft².

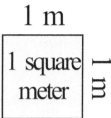

This is one square meter, or 1 m².

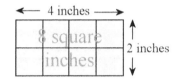

We need 8 square inches to cover this rectangle. So, its area is 8 square inches. We abbreviate this as 8 sq. in. or 8 in².

Again, use *multiplication* to find the area:

4 inches × 2 inches = 8 square inches

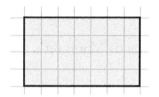

If no particular unit of length is given for the sides of a rectangle, we just use the word "unit."

The sides are 7 and 4 units, and the area is 28 *square units*.

3. Find the areas of the rectangles. Be very careful about the unit you need to use, whether square centimeters (cm²), square meters (m²), square inches (in²), or square feet (ft²).

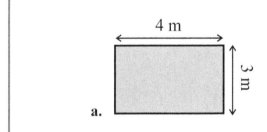

a.

A = _____

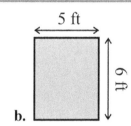

b.

A = _____

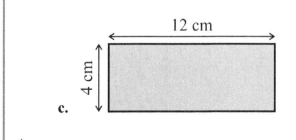

c.

A = _____

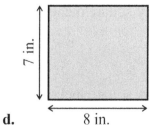

d.

A = _____

71

4. Find the area of this children's playground.

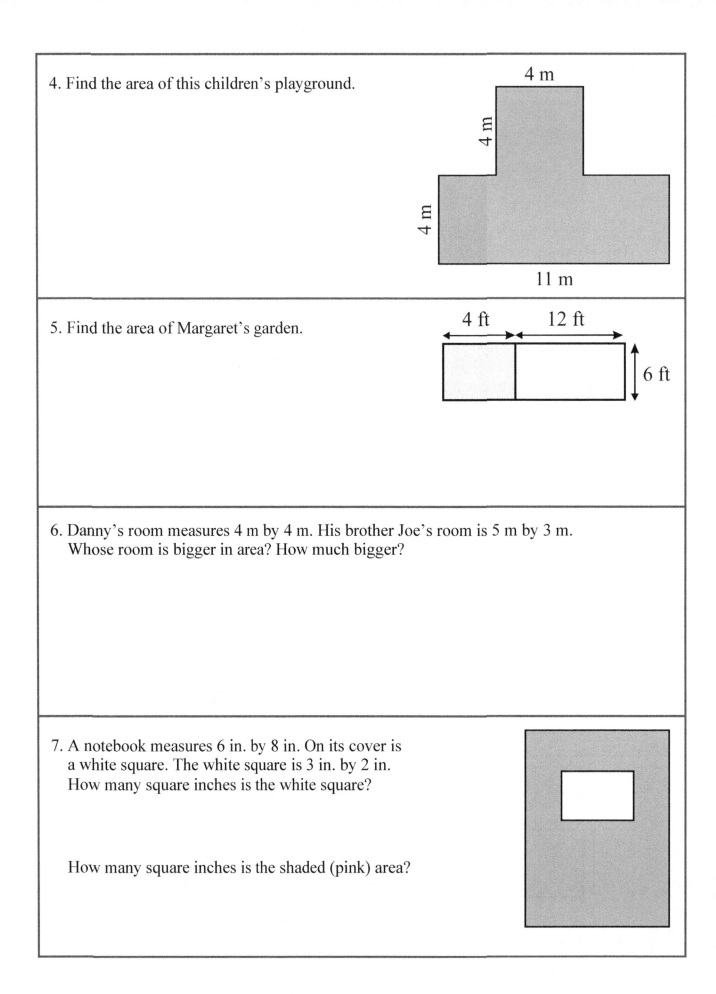

5. Find the area of Margaret's garden.

6. Danny's room measures 4 m by 4 m. His brother Joe's room is 5 m by 3 m. Whose room is bigger in area? How much bigger?

7. A notebook measures 6 in. by 8 in. On its cover is a white square. The white square is 3 in. by 2 in. How many square inches is the white square?

How many square inches is the shaded (pink) area?

Area and Perimeter Problems

Sometimes it's easy to confuse perimeter and area.

- AREA has to do with <u>covering the shape with squares</u>. Your answer will be in square centimeters, square inches, square feet, square meters, or just square units.

- PERIMETER has to do with "going all the way around." Your answer will be in some unit of length, such as centimeters, meters, inches, or feet.

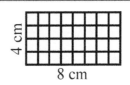

Area: 4 cm × 8 cm = 32 cm².

Perimeter: 4 cm + 8 cm + 4 cm + 8 cm = 24 cm

1. Find the area and perimeter of the rectangles.

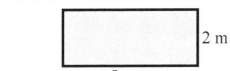

a. 5 m, 2 m

Perimeter = _____

Area = _____

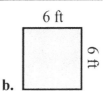

b. 6 ft, 6 ft

Perimeter = _____

Area = _____

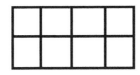

c. 4 in. wide, 2 in. tall

Perimeter = _____

Area = _____

d. A square with 3 cm sides

Perimeter = _____

Area = _____

2. Find the area and perimeter of this shape. Notice that one side length is not given. You need to figure that out.

Area:

Perimeter:

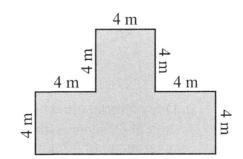

3. Find the area and perimeter of this shape.
   Notice that one side length is not given.
   You need to figure that out.

   Area:

   Perimeter:

4. This is a two-part lawn.

   a. Find the areas of the two parts.

   _____ and _____

   b. Find the total area.

   c. Find the perimeter.

5. Find the total area of this rectangle,
   and also the area of each little part.

   Area of each part:

   Total area:

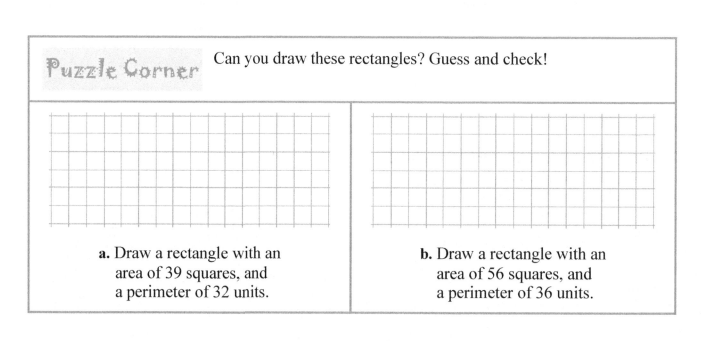

Puzzle Corner Can you draw these rectangles? Guess and check!

a. Draw a rectangle with an area of 39 squares, and a perimeter of 32 units.

b. Draw a rectangle with an area of 56 squares, and a perimeter of 36 units.

More Area and Perimeter Problems

1. **a.** Find the area for each part.

   _____ and _____

   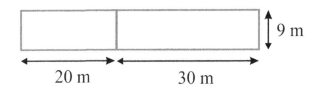

   b. Find the total area.

   c. Find the perimeter.

2. Make rectangles that have an area of 24 square units.

   Draw them in the grid. Write their side lengths in the table. One is already given.

	first side	second side	area
Rectangle 1	2 units	12 units	24 square units
Rectangle 2			24 square units
Rectangle 3			24 square units

3. For each rectangle you made in #2, calculate its perimeter.

	one side	second side	area	perimeter
Rectangle 1	2 units	12 units	24 square units	units
Rectangle 2			24 square units	
Rectangle 3			24 square units	

4. Make rectangles that have a perimeter of 20 units.
   Hint: the two different side lengths add up to half of the perimeter.

   Draw them in the grid.
   Write their side lengths in the table. One is already given.

	first side	second side	perimeter
Rectangle 1	2 units	8 units	20 units
Rectangle 2			20 units
Rectangle 3			20 units

5. For each rectangle you made in #4, calculate its area.

	first side	second side	perimeter	area
Rectangle 1	2 units	8 units	20 units	square units
Rectangle 2			20 units	
Rectangle 3			20 units	

6. The image illustrates Jane's garden.

   a. Find the area of each part.

   _____ and _____

   b. Find the total area.

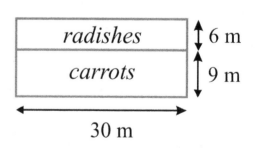

   c. Find the perimeter.

7. Draw and fill in.

a. Write a number sentence using the area of this two-part rectangle.

___ × (___ + ___) = ___ × ___ + ___ × ___

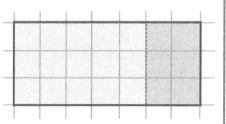

b. Draw a two-part rectangle to illustrate this number sentence.

4 × (3 + 5) = 4 × 3 + 4 × 5

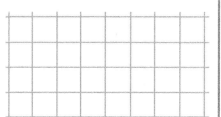

c. Fill in the missing parts, and then draw a two-part rectangle to illustrate this number sentence.

2 × (5 + 2) = ___ × ___ + ___ × ___

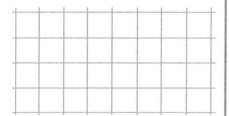

d. Fill in the missing parts, and then draw a two-part rectangle to illustrate this number sentence.

___ × (___ + ___) = 3 × 2 + 3 × 1

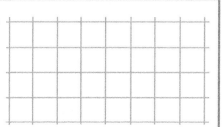

Puzzle Corner

a. Write a number sentence using the area of this two-part rectangle.

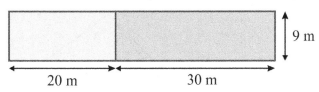

___ × (___ + ___) = ___ × ___ + ___ × ___

b. Sketch a rectangle to match 20 × (3 + 7) and find its area.

Solids

You can make paper models of these solids with the help of the printable cutouts provided (see introduction).

Solids are shapes that do not exist just on paper: you can fill them with something, such as water or stones. We say they are three-dimensional shapes.

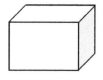  

A rectangular prism. We also call it a box. Its faces are rectangles.

A cube: all of its edges are of the same length.

A cylinder

A cone

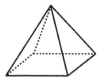 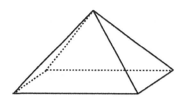

A square pyramid: its base (bottom) is a square.

A rectangular pyramid has a rectangle as its base.

This is a pyramid, too. It has a triangle at the bottom, so it has a special name: **a tetrahedron**.

Let's also study the parts of solids: **faces, edges, and vertices.**

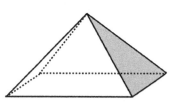

A face is a flat side with area.

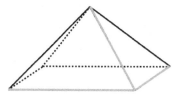

An **edge** is a boundary "line" for the face.

A **vertex** (pl. vertices) is a place where edges meet, like a corner.

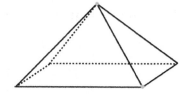

1. Name the shapes, and find how many <u>E</u>dges, <u>V</u>ertices, and <u>F</u>aces they have.

Shape	Name	E	V	F
a.				
b.				
c.				

2. **a.** Name (and find) some items that are in the shape of a cone.

   b. Name (and find) some items that are in the shape of a cylinder.

3. Where do each of these figures belong in the table? Try to do this exercise without checking back!

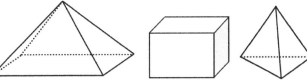

Name of Shape	Faces	Edges	Vertices
	6	12	8
	5	8	5
	4	6	4

Mixed Review Chapter 7

1. Write the time the clock shows.

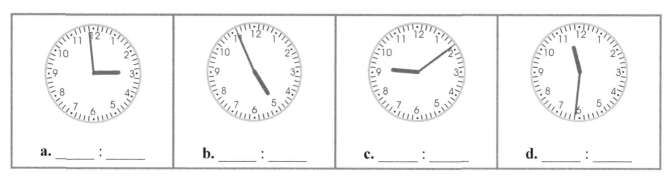

a. ____ : ____   b. ____ : ____   c. ____ : ____   d. ____ : ____

2. Subtract.

a. 456
   −163
   ____

b. 721
   −255
   ____

c. 4802
   −2316
   ____

d. 3700
   −1538
   ____

3. Below the addition, write a matching subtraction problem so that the numbers in the boxes are the same. Use mental math.

a. 99 + ☐ = 145

____ − ____ = ☐

b. 34 + ☐ = 76

____ − ____ = ☐

4. Compare. Write < , > , or = in the box.

a. 800 + 4000 ☐ 5000 + 400 + 80

b. 3000 + 60 + 5 ☐ 365

c. 20 + 8000 ☐ 4 + 8000 + 200

d. 400 + 9000 + 8 ☐ 80 + 900 + 8

e. 1 + 500 + 3000 ☐ 50 + 3000 + 900 + 9

f. 200 + 6000 + 40 + 7 ☐ 600 + 7000 + 2

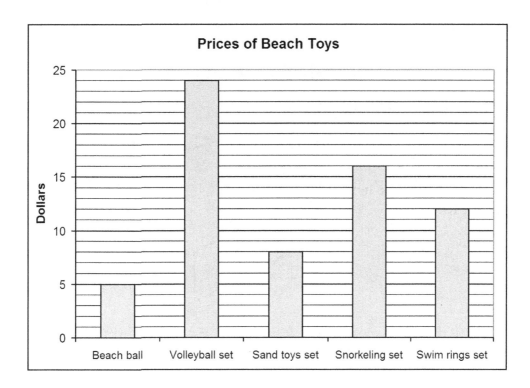

5. **a.** How much does the volleyball set cost?

   b. How much more does the snorkeling set cost than the swim rings set?

   c. How much do the sand toys and beach ball cost together?

   d. What is the total cost if you buy the two cheapest items?

6. Find the change.

a. A notebook costs $2.55. You give $3. Change: $_____	**b.** A book costs $5.88. You give $10. Change: $_____	**c.** A toy costs $6.70. You give $10. Change: $_____

7. Add parentheses to each equation to make it true.

a. $10 - 40 - 30 = 0$	**b.** $4 + 5 \times 2 - 1 = 17$	**c.** $5 \times 7 - 3 - 1 = 19$

Geometry Review

1. **a.** Find the rhombi among these figures.

   b. Find quadrilaterals that are neither rectangles nor rhombi.

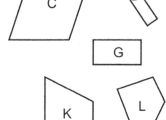
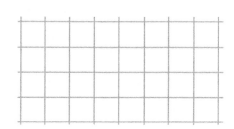

2. Draw a quadrilateral that is not a rectangle.

3. Fill in.

a. Write a multiplication for the area of this figure. 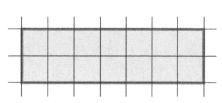 ___ units × ___ units = ___ square units	**b.** Draw a rectangle that has the area shown by the multiplication. 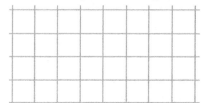 4 × 5 = 20 square units

4. Find the perimeter and area of this rectangle. Use a centimeter ruler.

   Area:

   Perimeter:

5. Find the area and perimeter of these figures.

a. 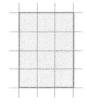 Area:

Perimeter:

b. 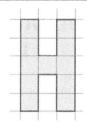 Area:

Perimeter:

6. Write a multiplication *and* addition for the areas of these figures.

a.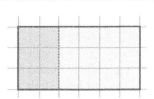

A = _____

b.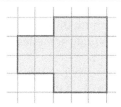

A = _____

7. Multiply using the shortcut.

a. 7 × 70 = _____   b. 6 × 80 = _____   c. 40 × 7 = _____

8. Find the total area of this rectangle, and the area of each part.

Area of each part:

Total area:

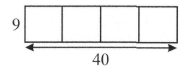

9. Draw and fill in.

a. Fill in the missing parts, and then draw a two-part rectangle to illustrate this number sentence.

3 × (5 + 1) = ___ × ___ + ___ × ___

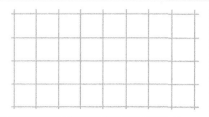

b. Fill in the missing parts, and then draw a two-part rectangle to illustrate this number sentence.

___ × (___ + ___) = 4 × 2 + 4 × 3

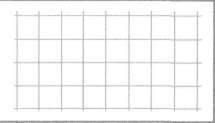

Chapter 8: Measuring
Introduction

In this chapter we delve into both customary and metric measuring units.

Note: If you have the downloadable version of this book (PDF file), you need to print the pages as 100%, not "shrink to fit," "print to fit," or similar. If you print "shrink to fit", some exercises about measuring in inches and centimeters will not come out right, but will be "shrunk" compared to reality.

First, students learn about units of length. We start by measuring to the nearest quarter of an inch. Since most rulers measure to the eighth or sixteenth part of an inch, it is helpful to cut out a ruler from the lesson that only has tick marks for every fourth of an inch, and tape that onto an existing ruler. If your student has trouble with the fractions, consider studying some lessons from chapter 10 (Fractions) first.

Next, students measure using centimeters and millimeters. They also create line plots from measurement data where the horizontal scale is marked off in quarters of an inch. The next two lessons help students become familiar with feet, yards, miles, meters, and kilometers—the units for measuring medium and long distances.

Then it is time to measure weight. First, we deal with pounds and ounces, and next, with grams and kilograms. It is very helpful if you can use a kitchen scale for these lessons.

Lastly, we study liquid volume, first of all with customary units (cup, pint, quart, and gallon) and then with metric units (liter and milliliter).

Many of the lessons in this chapter also have an optional section about conversions between measuring units, such as changing meters into centimeters, or feet into inches. Converting between units is beyond the Common Core standards for third grade (it is actually included in the 4th and 5th grade standards), but I have included some easy conversion problems here, because I feel many third graders are ready for them.

We all use various measuring units in our everyday lives, and using them is the key to remembering what they are, how big they are, and what the conversion factors are. People in the United States do not use the metric system a lot, while people in other countries do not use the customary system. The units that you do not use are likely to be forgotten. So encourage the student(s) to have free play time with measuring devices such as scales, measuring cups, measuring tapes, and rulers.

The Lessons

	page	span
Measuring to the Nearest Fourth-Inch	87	*4 pages*
Centimeters and Millimeters	91	*4 pages*
Line Plots and More Measuring	95	*3 pages*
Feet, Yards, and Miles	98	*2 pages*
Meters and Kilometers	100	*2 pages*
Pounds and Ounces	102	*4 pages*
Grams and Kilograms	106	*4 pages*
Cups, Pints, Quarts, and Gallons	110	*3 pages*
Milliliters and Liters	113	*2 pages*
Mixed Review Chapter 8	115	*2 pages*
Review Chapter 8	117	*2 pages*

Helpful Resources on the Internet

You can use these free online resources to supplement the "bookwork" as you see fit.

GENERAL

Measures
An online activity about metric measuring units and how to read scales, a measuring cup, and a ruler.
Note: you will need to use the British spellings "centimetres" and "millilitres" in the activity.
http://flash.topmarks.co.uk/674

Conversion Quizzes - ThatQuiz.org
Create customizable quizzes about conversions between measuring units.
http://www.thatquiz.org/tq-n/science/metric-system/

Reading Scales
Illustrate how to read a variety of measuring devices, such as scales, measuring cup, thermometer. You can generate examples using different scales on different devices.
http://www.teacherled.com/resources/dials/dialsload.html

MEASURING LENGTH

The Ruler Game
Click on the given measurement on a ruler. You can choose to practice whole inches, halves, 1/4, 1/8, or 1/16 parts of an inch.
http://www.rulergame.net/

Measure It!
Practice measuring lines with either centimeters or inches. Multiple choice questions.
https://www.funbrain.com/games/measure-it

Reading a Tape Measure Worksheets
Generate printable worksheets - you can choose to which accuracy to measure: inches, or inches and feet.
http://themathworksheetsite.com/read_tape.html

Reading a Metric Ruler
This page has illustrated instructions and then a short practice exercise.
http://www.texasgateway.org/node/3970

Online Measurement Game
Drag the pointer to the position on the ruler that corresponds to the correct answer.
http://www.bigiqkids.com/MeasurementGame.shtml

Measuring - Find Lengths with a Ruler
Drag the ruler to measure the length of the given lines. Choose "Tenths" for this grade level, then enter the length using a decimal, such as 0.3 cm.
http://www.abcya.com/measuring.htm

Metric Length Matching
Match the correct conversions.
http://www.sheppardsoftware.com/mathgames/measurement/MeasurementMeters.htm

LINE PLOTS

Data Analysis: Line Plots
First, play a game. Then, make a line plot using the game scores.
http://www.k5learning.com/sample-lessons/grade-3-data-analysis

Solve Problems with Line Plots
Answer questions using line plots and data sets.
https://www.khanacademy.org/math/early-math/cc-early-math-measure-data-topic/cc-early-math-line-plots/e/solving-problems-with-line-plots-1

MEASURING WEIGHT/MASS

Interactive Measuring Scales
Add weights to the scales and choose to show or hide the total weight.
http://www.taw.org.uk/lic/itp/itps/measuringScales_1_8.swf

Scales Reader
Simple online practice of reading the scales. Choose "up to 500 g" or "up to 1 kg" for this level.
http://www.ictgames.com/weight.html

Mostly Postie!
Choose "grams". Place a letter on the scale, and enter the reading, and click "check."
http://www.ictgames.com/mostlyPostie.html

Ounce or Pound
Click and drag to show which unit you would use to weigh the object.
http://www.harcourtschool.com/activity/ounces_pounds/

Get the Weight
Estimate the weight of the items that are placed on the balance scale. The longer you hold down the mouse button, the bigger your estimate of its weight. Available both for customary and metric units.
http://www.mathnook.com/math/get-the-weight-standard.html

http://www.mathnook.com/math/get-the-weight-metric.html

Best Measure
Match each thing with its best estimated weight.
http://www.sheppardsoftware.com/mathgames/measurement/BestMeasure2.htm

Measurement Game for Kids
Measure the length and weight of various parcels using the interactive scales and ruler, so you can give them a stamp with the correct postage rate. Uses grams and centimeters.
http://www.kidsmathgamesonline.com/geometry/measurement.html

VOLUME/CAPACITY

Gallon Bot or Gallon Man
This is a graphical creation that allows students to better visualize the customary units of volume.
https://www.superteacherworksheets.com/pz-gallon-man.html

Taking Measures Capacity Game
Click on the object on the table that best matches the measure or object at the top of the screen.
http://www.bbc.co.uk/skillswise/game/ma23capa-game-taking-measures-capacity

Measuring to the Nearest Fourth-Inch

This ruler measures in inches. You can see three lines between each two numbers on the ruler. Those three lines divide each inch *into four parts*. The parts are *fourth parts* or *quarters* of an inch. We have marked those quarters with fractions.

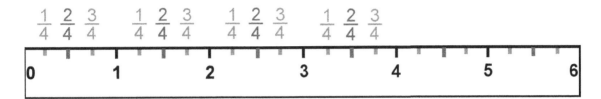

The 2/4 mark is also the 1/2 mark. We normally use 1/2 instead of 2/4.

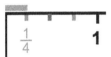   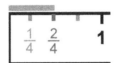   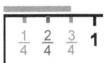

This line is 1/4 inch long.   This line is 2/4 inch long. It is also 1/2 inch long.   This line is 3/4 inch long.

If a line reaches to the 1/4-inch mark after the number 1, then the line is 1 inch *and* 1/4 inch long. But when writing it, we omit the "and" and write: The line is 1 1/4 inches long.

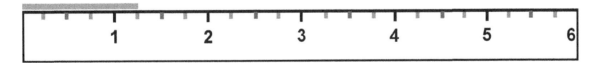

If a line reaches the 3/4-inch mark after the number 2, then the line is 2 inches *and* 3/4 inch long, but we write it as 2 3/4 inches long.

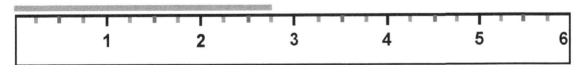

This line is 3 1/2 inches long.

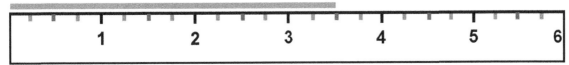

1. Measure the lines using the ruler.

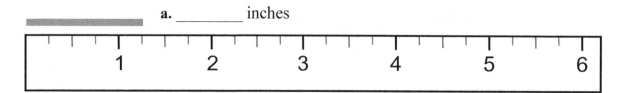

a. _____ inches

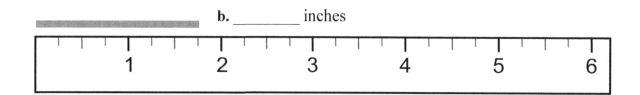

b. _____ inches

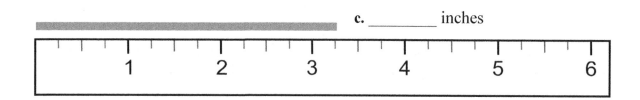

c. _____ inches

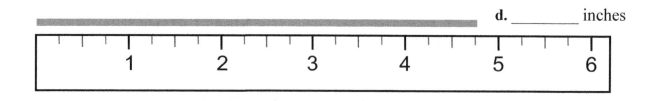

d. _____ inches

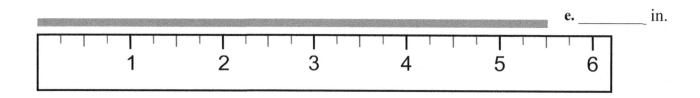

e. _____ in.

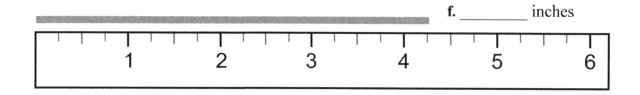

f. _____ inches

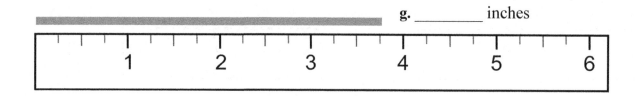

g. _____ inches

2. Draw lines using a ruler. Your own ruler may have many more little lines between the whole inch marks. If you find your own ruler confusing, you can cut out one of the rulers from the previous pages, and use that. Glue it on cardboard, or place it on top of your ruler.

   a. 4 1/2 inches long

   b. 2 1/4 inches long

   c. 5 1/4 inches long

   d. 4 3/4 inches long

This line is not exactly 3/4 inch long, and not exactly 1 inch long, but its length is between those two. The endpoint of the line is closer to the 3/4-inch mark than it is to the 1-inch mark. We say the line is *about* 3/4 inch long, or *approximately* 3/4 inch long.

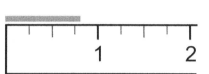

3. Measure items using the ruler that has the 1/4, 1/2, and 3/4 inch marks (quarters of an inch). If the item is not exactly as long as the markers on the ruler show, choose the nearest mark as the length, and write "about 5 1/4 inches," etc.

Item	Length

This page is optional.

Let's use the plus sign "+" to mean that we place two lines end-to-end.

1/4 inch + 3/4 inch = 1 inch

Here, the second line "covers" three short 1/4 inch segments, so it is 3/4 inch long.

1/4 inch + 1 inch = 1 1/4 inches

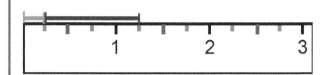

Here, the second line "covers" four short 1/4 inch segments, so that is why it is 1 inch long.

4. Draw another line after the 1/4-inch line. Add the lengths and find the total length.

a. 1/4 inch + 1/4 inch = _____ inches

b. 1/4 inch + 1/2 inch = _____ inches

c. 1/4 inch + 1 1/4 inch = _____ inches

d. 1/4 inch + 2 inch = _____ inches

5. Work out these "line additions". You can use the ruler below to help. Or, you can draw the lines.

a. 1/4 in. + 1/4 in. = _____

1 1/4 in. + 1/4 in. = _____

b. 1/4 in. + 3/4 in. = _____

4 1/4 in. + 1/4 in. = _____

c. 5 1/4 in. + 3/4 in. = _____

7 3/4 in. + 1/4 in. = _____

d. 1/2 in. + 1/4 in. = _____

2 1/2 in. + 1/4 in. = _____

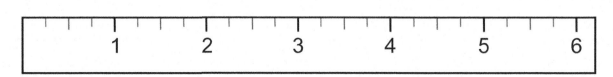

Centimeters and Millimeters

This ruler measures in centimeters. The numbers signify whole centimeters. All the shorter lines between those are for *millimeters*.

The distance from one short line to the next line is *1 millimeter*. We write 1 mm. Millimeters are very tiny!

The distance between these two is 1 mm.

Look at the ruler: **there are 10 millimeters in each centimeter.**

Measuring lines: First see how many whole centimeters long the line is. Then count how many little millimeter-lines beyond that it reaches.

This line is 2 cm 3 mm long. At the same time, it is 23 mm long. Why?

Each centimeter is 10 mm, so 2 cm is 20 mm. That means 2 cm 3 mm makes 23 mm in total.

This line is 4 cm 8 mm long. At the same time, it is 48 mm long.

1. Measure the lines using the ruler, first in whole centimeters and millimeters. Then write their lengths using millimeters only.

   a. _____ cm _____ mm = _____ mm

   b. _____ cm _____ mm = _____ mm

[ruler image]

c. _____ cm _____ mm = _____ mm

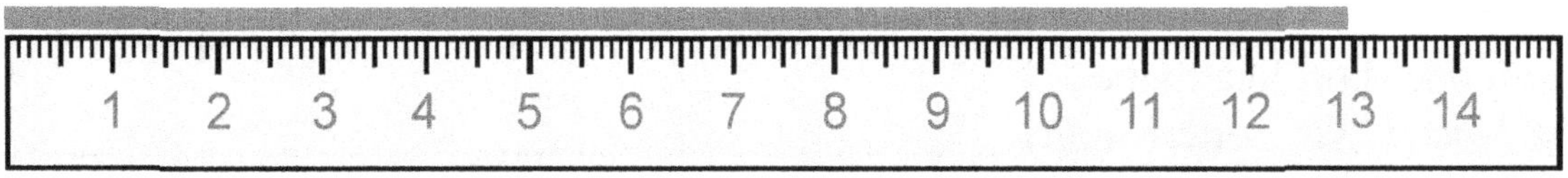

[ruler image]

d. _____ cm _____ mm = _____ mm

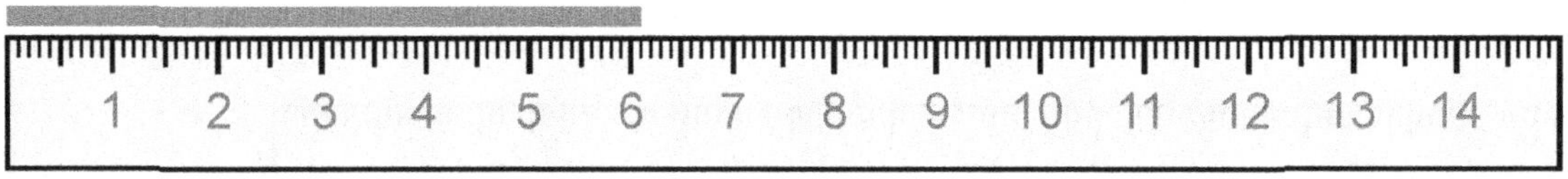

[ruler image]

e. _____ cm _____ mm = _____ mm

[ruler image]

f. _____ cm _____ mm = _____ mm

2. Draw lines using a ruler.

   a. 7 cm 8 mm

   b. 10 cm 5 mm

   c. 14 mm

   d. 55 mm

   e. 126 mm

3. Measure items you can find at home, using a centimeter-millimeter ruler.
   If the item is not exactly as long as the markers on the ruler, choose the nearest mark.

Item	Length

4. Change between centimeters and millimeters.

a.	b.	c.
1 cm = _____ mm	1 cm 1 mm = __11__ mm	4 cm 5 mm = _____ mm
2 cm = _____ mm	1 cm 2 mm = _____ mm	2 cm 5 mm = _____ mm
5 cm = _____ mm	1 cm 8 mm = _____ mm	7 cm 8 mm = _____ mm
8 cm = _____ mm	2 cm 3 mm = _____ mm	10 cm 4 mm = _____ mm

5. Change between millimeters and centimeters.

a.	b.	c.
70 mm = _____ cm	12 mm = ___ cm ___ mm	89 mm = ___ cm ___ mm
100 mm = _____ cm	45 mm = ___ cm ___ mm	102 mm = ___ cm ___ mm

6. Measure the sides of this triangle in millimeters.

Side AB _____ mm

Side BC _____ mm

Side CA _____ mm

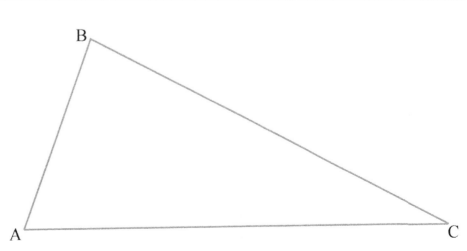

7. Find the perimeter of the triangle in the previous exercise.

8. Draw the third side of this triangle.
   Then measure its sides.
   Lastly, find its perimeter in millimeters.

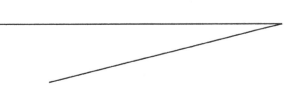

The first arrow is 4 cm. The second arrow is 1 cm 8 mm. How long are they together? Add, giving your answer in millimeters. 4 cm + 1 cm 8 mm = 5 cm 8 mm = 58 mm
Add centimeters with centimeters, and millimeters with millimeters. Remember that 10 millimeters makes 1 centimeter. 9 mm + 6 cm + 2 mm = 6 cm 11 mm = 7 cm 1 mm = 71 mm
If you have both millimeters and centimeters, change the centimeters to millimeters first: 84 mm + <u>3 cm</u> + 9 mm = 84 mm + <u>30 mm</u> + 9 mm = 123 mm (which is also 12 cm 3 mm)

9. Work out these "line additions." Give your answers in millimeters.

a. 1 cm 5 mm + 5 mm	b. 28 mm + 7 cm
c. 5 mm + 5 cm 8 mm	d. 2 cm 4 mm + 4 cm 5 mm
e. 52 mm + 2 cm 4 mm	f. 6 cm + 8 mm + 17 mm
g. 9 mm + 17 mm + 2 cm	h. 139 mm + 50 cm + 2 mm

Line Plots and More Measuring

1. Amanda measured the length of some of her colored pencils. She recorded her results in a **line plot** below. For each pencil, she put an "x" mark above the number line to show how many inches long it was.

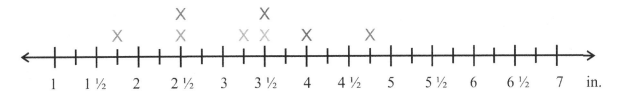

   Look carefully at the line plot, and find the X-marks:

   - There is one pencil that is 4 inches long.
   - There are two pencils that are 3 1/2 inches long
   - There are two pencils that are 2 1/2 inches long.

   a. There is one pencil whose X-mark is between 3 and 3 1/2 inches. How long is it?

   b. How long is the pencil whose X-mark is between 4 1/2 and 5?

   c. How long is the pencil whose X-mark is between 1 1/2 and 2?

2. Draw three dots in the space on the right and join them to get a triangle.

   Measure its sides to the nearest quarter inch. Write the measurement next to each side.

   If you can, figure out the *perimeter*. (all the way around the shape)

   It is _____ inches.

You can cut out the ruler below, and glue it on cardboard or on top of your ruler.

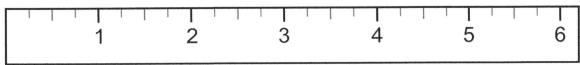

3. Measure several pencils of different lengths to the nearest quarter inch.
   Write the lengths below.

   _____ in, _____ in, _____ in, _____ in, _____ in.

   _____ in, _____ in, _____ in, _____ in, _____ in.

   _____ in, _____ in, _____ in, _____ in, _____ in.

   Now, make a line plot about your pencils. Write an "X" mark for each pencil.

   ←|—|—|—|—|—|—|—|—|—|—|—|—|—|→
   1 1½ 2 2½ 3 3½ 4 4½ 5 5½ 6 6½ 7 in.

   a. If you take your two longest pencils and put them end-to-end,
   how long is your line of pencils?

   It is _____ in.   (You can measure to check your answer!)

   b. If you take your two shortest pencils and put them end-to-end,
   how long is your line of pencils?

   It is _____ in.   (You can measure to check your answer!)

4. Measure all the sides of this shape in centimeters and millimeters.
   Can you figure out the perimeter? It can be a little tricky, but try!

   Side AB _____ cm _____ mm

   Side BC _____ cm _____ mm

   Side CD _____ cm _____ mm

   Side DA _____ cm _____ mm

   Perimeter _____ cm _____ mm

5. Measure some things in your classroom or at home *two times*. First measure them in inches, to the nearest quarter-inch. Then measure them in centimeters and millimeters. Each time, GUESS before you actually measure. Write your results in the table below.

Item	GUESS in inches	LENGTH in inches	GUESS in cm / mm	Length in cm / mm
	___ in	___ in	___ cm ___ mm	___ cm ___ mm
	___ in	___ in	___ cm ___ mm	___ cm ___ mm
	___ in	___ in	___ cm ___ mm	___ cm ___ mm
	___ in	___ in	___ cm ___ mm	___ cm ___ mm
	___ in	___ in	___ cm ___ mm	___ cm ___ mm

6. Measure a collection of similar items to the *nearest* quarter inch. For example, you can measure some spoons, lots of stuffed animals, or the width of lots of books. Or, ask some people to draw a line 6 inches long without using a ruler (in other words, guess and draw it), and then measure their lines and check who guessed the closest.

(You do not have to find as many items as there are empty lines below.)

_____ in, _____ in, _____ in, _____ in, _____ in.

_____ in, _____ in, _____ in, _____ in, _____ in.

_____ in, _____ in, _____ in, _____ in, _____ in.

_____ in, _____ in, _____ in, _____ in, _____ in.

Now, make a line plot. Write an "X" mark for each item.

←┼┼┼→ in.
2 2½ 3 3½ 4 4½ 5 5½ 6 6½ 7 7½ 8 8½ 9 9½ 10 10½ 11 11½ 12

Feet, Yards, and Miles

Feet and yards are used to measure the length of medium-size objects and distances. A foot is abbreviated with "ft" and a yard is abbreviated with "yd".

Three feet make one yard.
3 ft = 1 yd.

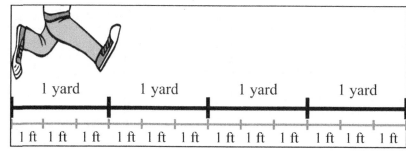

A *mile* is used to measure long distances, such as between towns or countries.
One mile is 5,280 feet. An adult can typically walk 1 mile in about 15-20 minutes.

1. Outside, using a measuring tape, mark the distances of 1 yard, 2 yards, 3 yards, and so on. Measure also, using feet: there are three feet in each yard.

    Take steps that are 1 foot long. That should be easy.

    How about steps that are 2 feet long each?

    Lastly, try to take steps 1 yard long (three feet). Can you?

2. Write or say these units in order from the smallest to the biggest:

    yard   mile   inch   foot

3. Use a tape measure to measure lengths of some objects and distances in feet and inches.

Item	How long
	_____ ft _____ in.

4. Fill in the blanks, using the units in., ft., or mi.

   a. Mark drove his car 15 _____ .

   b. The table is 24 _____ tall.

   c. Annie's house is 32 _____ long.

   d. The pen is 5 _____ long.

   e. Mr. Green is 6 _____ tall.

   f. Matt jogged 3 _____ .

5. Find the area and perimeter of a rectangular yard that is 30 ft by 10 ft.
   Hint: Make a sketch (a picture) of it.

This section is optional.

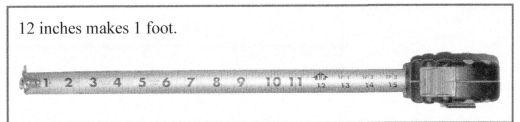

12 inches makes 1 foot.

6. Convert between feet and inches.

a. 1 ft = _____ in	**b.** 1 ft 2 in = _____ in.	**c.** 2 ft 4 in = _____ in
3 ft = _____ in	1 ft 8 in = _____ in.	2 ft 6 in = _____ in
5 ft = _____ in	1 ft 11 in = _____ in.	3 ft 3 in = _____ in

7. Emma is 4 ft 2 in. tall. How tall is she in inches?

8. Mary is 3 ft 9 in. and Rebecca is 48 inches tall.
   Who is taller? How much taller?

9. Alex put three pencils end-to-end that each were 8 inches long.
   How many feet long is his line of pencils?

10. Find the perimeter of a rectangle with one side 2 ft 2 in
    and the other side 3 ft 8 in.

Meters and Kilometers

Besides feet, yards, and miles, we can also use millimeters (mm), centimeters (cm), meters (m), and kilometers (km) to measure length.

Notice how all of those units have the word *meter* in them. These units form a part of the *metric system* of measuring units.

kilometer for long distances
meter for medium-sized objects and distances
millimeter **centimeter** } for small objects

1. Outside, mark the distances of 1 meter, 2 meters, 3 meters, and so on, using a measuring tape. Try to take steps 1 meter long. Can you?

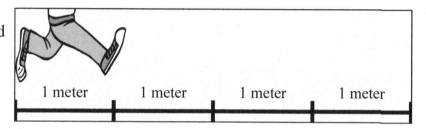

If you can't, try to take small steps so that two steps would be exactly 1 meter.

Notice: one meter is very close to one yard, but a little longer.

2. Use a tape measure to measure lengths of objects and distances in meters and centimeters

Item	How long
	_____ m _____ cm.

3. Fill in the blanks, using the units "cm", "km", "mm", and "m".

   a. The Jacksons' living room is 4 ____ wide.

   b. A moth was 38 ____ wide.

   c. Dad is 178 ____ tall.

   d. It is about 3 ____ to the nearest library.

   e. The window was about 1 ____ wide.

   f. The book was 25 ____ long.

4. Use a measuring tape or a tape measure, and measure how tall you and some other people are in meters and centimeters.

Person	How tall
	1 m _____ cm.

5. Write or say these units in order from smallest to greatest, using their full names:

   m   cm   km   mm

   _____

This section is optional and is beyond the Common Core Standards.

$$1 \text{ meter} = 100 \text{ cm}$$

6. Convert between meters and centimeters.

a. 1 m = _____ cm	**b.** 1 m 20 cm = _____ cm	**c.** 5 m 85 cm = _____ cm
2 m = _____ cm	1 m 14 cm = _____ cm	2 m 17 cm = _____ cm
5 m = _____ cm	1 m 58 cm = _____ cm	3 m 8 cm = _____ cm

7. One pillow is 40 cm long. If you put five such pillows end-to-end, how many meters long is your line of pillows?

8. Ellie is 162 cm tall, and Meredith is 1 m 55 cm tall. Who is taller? How much taller?

9. A sandbox is 1 m 40 cm by 1 m 40 cm (a square). Find its perimeter.

Pounds and Ounces

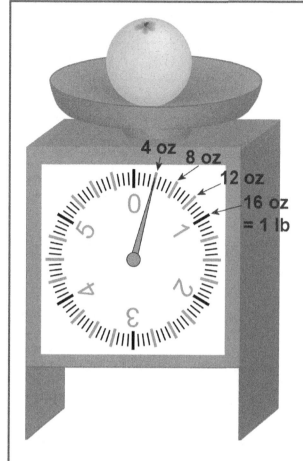

This is a kitchen scale that measures in *pounds* and *ounces*. We use it to measure the weight of small items.

This scale can measure items from 0 to 6 pounds. The numbers 0, 1, 2, 3, 4, and 5 refer to whole pounds. Number 6 is not marked, but if the pointer went all the way around one time and was pointing to 0, it would actually mean 6 pounds.

Each pound is divided into 16 ounces. So, an ounce is a small unit of weight.

In between the whole pounds are lines to mark the ounces. Some are longer and thicker (red), and some are shorter.

The thicker lines mark the 4-ounce, 8-ounce, and 12-ounce points, and the shorter lines mark the individual ounces.

A pound or pounds is abbreviated with "lb." An ounce or ounces is abbreviated with "oz." So, 2 lb 13 oz means 2 pounds 13 ounces.

The orange weighs 4 ounces.

1. Write the pounds and the ounces the scale is showing.

a. _____ lb _____ oz **b.** _____ lb _____ oz **c.** _____ lb _____ oz

2. Write the pounds and the ounces the scale is showing.

a. _____ lb _____ oz   b. _____ lb _____ oz   c. _____ lb _____ oz

  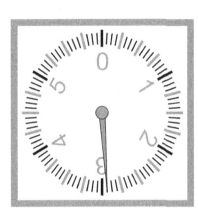

d. _____ lb _____ oz   e. _____ lb _____ oz   f. _____ lb _____ oz

g. _____ lb _____ oz   h. _____ lb _____ oz   i. _____ lb _____ oz

3. Weigh light items with a kitchen scale. Write your results here.

Item	Weight
	_____ lb _____ oz

4. Weigh things and people with a bathroom scale that uses pounds. Write your results here. First, guess how much the thing or person weighs. Then weigh using the scales.

Thing/person	Guess	Weight
	_____ lb	_____ lb

5. At home, find food products that show the weight on the label, using ounces or pounds and ounces. Write the items in order from lightest to heaviest.

Item	Weight

6. Which is the best estimate of weight?

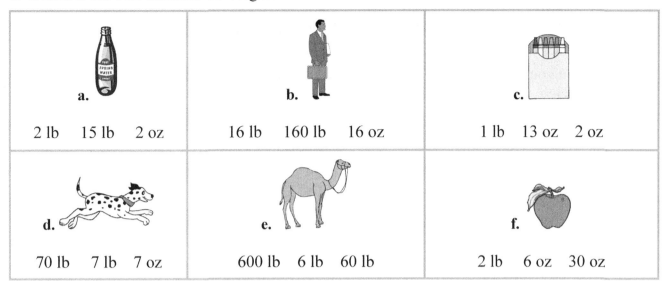

a.	b.	c.
2 lb 15 lb 2 oz	16 lb 160 lb 16 oz	1 lb 13 oz 2 oz
d.	e.	f.
70 lb 7 lb 7 oz	600 lb 6 lb 60 lb	2 lb 6 oz 30 oz

7. Fill in the blanks with a reasonable unit of weight (either lb or oz).

   a. A computer weighs 3_____. **b.** A newborn baby weighed 8_____.

   c. Sam ate two bananas. Together they weighed 12 _____.

   d. Abby's cell phone weighs 3 _____. **e.** Matthew weighs 170 _____.

The following problems are optional.

In the following problems, use the fact that 1 lb = 16 oz.

8. Convert between pounds and ounces.

a. 2 lb = _____ oz	**b.** 1 lb 1 oz = _____ oz	**c.** 2 lb 4 oz = _____ oz
3 lb = _____ oz	1 lb 7 oz = _____ oz	3 lb 9 oz = _____ oz
4 lb = _____ oz	2 lb 11 oz = _____ oz	5 lb 4 oz = _____ oz

9. A label on a big tuna can says: "Net weight 1 lb. Drained weight 11 oz."
   How much does the liquid in the can weigh?

10. Mary sent thank you letters to people who had attended a birthday party for her fiftieth birthday. Each letter weighed two ounces.

   a. How many 2 oz letters will weigh a total of 1 pound?

   b. She sent 15 letters. What was their total weight, in pounds/ounces?

Grams and Kilograms

In this lesson you will need a bathroom scale that measures weight in *kilograms* (abbreviated kg).

The scale on the right is showing 22 kg.

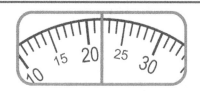

You will also need a kitchen scale that measures in grams. A *gram* is a very small unit of weight. A gram is abbreviated with "g".

A thousand grams make one *kilogram* (1 kg): 1,000 g = 1 kg.

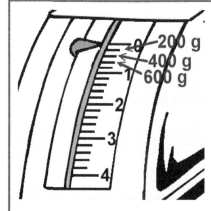

Look carefully at the kitchen scale. The numbers 0, 1, 2, 3, and 4 on this scale refer to the whole kilograms.

In between each two numbers there are four little lines. They divide each kilogram into *five* parts. This means that each little line marks a 200-gram increment.
(200 g + 200 g + 200 g + 200 g + 200 g = 1,000 g = 1 kg.)

The first little line after the 0-kg mark means 200 g. The next little line means 400 g, the next one 600 g, and so on. Each time, one little line more means 200 g more.

1. Write the amount of kilograms and grams that the scales are showing.

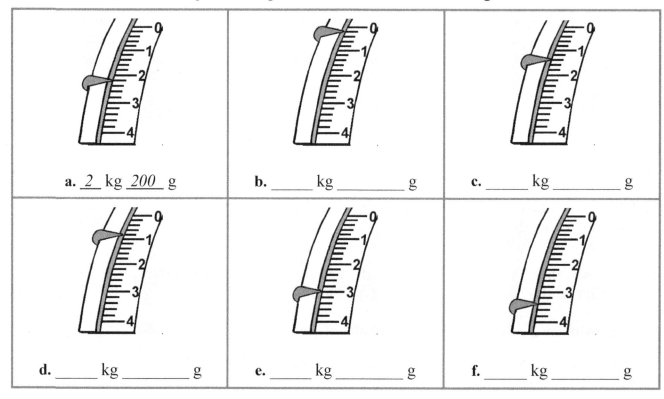

a. _2_ kg _200_ g   b. ____ kg ____ g   c. ____ kg ____ g

d. ____ kg ____ g   e. ____ kg ____ g   f. ____ kg ____ g

2. Weigh things and people with a bathroom scale that uses kilograms. Write your results here. First, guess how much the thing or person weighs. Then weigh using the scales.

Thing/person	Guess	Weight
	_____ kg	_____ kg

3. Measure light items with a kitchen scale that uses grams. Write your results here.

Item	Weight

4. At home, find food products or personal care products that show the weight on the label using grams. Write the items in order from the lightest to the heaviest.

Item	Weight
	_____ g

5. Which is the best estimate of weight?

a.	b.	c.	d.
500 g 5 g	70 kg 7 kg	1 kg 200 g	1 kg 150 g
e.	f. a CD	g. A car	h. an apple
30 g 300 g	15 g 300 g	100 kg 2,000 kg	1 kg 100 g
i. A bucket full of water	j. A spoonful of sand	k. a baby	l. flashlight
8 kg 50 kg	10 g 100 g	500 g 5 kg	300 g 2,000 g

6. Match the things and their weights.

An adult woman	55 kg
A puppy	1 kg
A pencil	500 g
A school book	25 kg
A magazine	50 g
A 9-year-old boy	150 g

7. Fill in the blanks with a reasonable unit of weight (either g or kg).

   a. Mom got a package in the mail that weighed 3 _____. It had books in it.

   b. Jane got a package in the mail that weighed 300 _____. It had a puzzle in it.

   c. Mark's dog weighs 30 _____.

   d. A cell phone weighs 300 _____.

   e. Mary bought 3 _____ of strawberries at the marketplace.

   f. Audrey weighs 60 _____.

   g. The teddy bear weighs 250 _____.

Convert between kilograms and grams *
(*This section is optional.*)

Just remember that 1 kg = 1000 grams, and use that when changing between kilograms and grams.

3 kg = 3,000 g

1 kg 500 g = 1,000 g + 500 g = 1,500 g

8. Convert between kilograms and grams.

a. 1 kg = _____ g	b. 1 kg 600 g = _____ g	c. 9 kg = _____ g
2 kg = _____ g	1 kg 80 g = _____ g	8 kg 600 g = _____ g
3 kg = _____ g	2 kg 450 g = _____ g	5 kg 8 g = _____ g
4 kg = _____ g	8 kg 394 g = _____ g	7 kg 41 g = _____ g

9. A t-shirt weighs 200 g. How many of those would weigh 1 kg?

10. A math book weighs 1 kg 300 g. An English book weighs 1 kg 700 g. How much do they weigh together?

11. Anne's school books weigh 800 g, 700 g, and 600 g.

    a. What is their total weight in grams?

    b. What is their total weight in kilograms and grams?

12. Marlene bought 2 kg 400 g of potatoes. She used 500 grams in soup. How much do the remaining potatoes weigh?

13. Greg has a post office box that allows him to receive 10 kg of mail each month. So far this month he has received packages that weighed 1 kg 500 g, 4 kg 800 g, and 2 kg.

    a. What is the total weight of the packages he has received this month?

    b. What is the total weight of mail he is still allowed to receive this month?

Cups, Pints, Quarts, and Gallons

Volume means *how much space* something takes.

A sandcastle takes a certain amount of space. A bottle of water takes space. A book takes space. But how much?

In this lesson you will learn how we measure the volume of water (or other liquids).

You will need

- water in a bucket or other big container
- a few food containers
- a coffee cup
- a drinking glass
- a quart jar
- a pint jar
- a 1-cup measuring cup

1. Fill the <u>pint</u> jar with water. Pour it all into the quart jar. Then fill the pint jar again and pour it into the quart jar. Is it now full (or close to full)?

   It should be. It takes _____ pints of water to fill 1 quart jar.

2. Pour out water from your full quart jar back into the pint jar until the pint jar is full.

   Is your quart jar now half full? (It should be.)

   How much water is left in the quart jar? _____ pint.

3. Find out how many times you need to fill the one-cup measuring cup with water and pour it into the pint jar until the pint jar is full.   _____ times.

   One pint is _____ cups.

4. Find out how many times you need to fill the one-cup measuring cup with water and pour it into the quart jar until the quart jar is full.   _____ times.

   One quart is _____ cups.

5. Find out if a coffee cup measures MORE or LESS than the a 1-cup measuring cup—or exactly 1 cup. Do the same with a drinking glass.

6. Find three different empty food containers. Measure water into them, and find out how many whole cups of water you can fit into them. If you can still fit a little more, write YES.

	how many whole cups	Can you fit a little more?
Container 1		
Container 2		
Container 3		

7. At the next supper or breakfast time, do a little experiment. Before eating, measure exactly one cup of the food you are going to eat and then put it on your plate. Will it fill you up? Is it too much or too little food?

_____

This section is optional.

• A quart is abbreviated with "qt".	5 qt means 5 quarts.
• A pint is abbreviated with "pt".	3 pt means 3 pints.
• A cup is abbreviated with "C".	2 C means 2 cups.

8. Fill in numbers on the blank lines. You will get help from your work on the previous page.

a. 1 qt = _____ pt	**b.** 1 qt = _____ C	**c.** 1 pt = _____ C

9. Circle the amount that holds more liquid volume. Circle both if they hold the same amount.

a. pt OR (cup)	**b.** pt OR (2 cups)	**c.** pt OR (2 cups)
d. pt OR qt	**e.** qt OR (2 cups)	**f.** pt pt OR qt

10. Fill in with the words cup, pint, or quart.

   a. Mary drank 2 _____s of tea at the party.

   b. Mom bought 1 _____ of yogurt for the four children.

   c. Ron was quite thirsty and so he drank a whole _____ of water.

   d. The large pitcher can hold 2 _____s of juice.

Gallons

One gallon is a large measure of volume. You use gallons when the liquid or other substance takes a lot of space, even more than a few quarts.

You might have heard about these items. Fill in some more items that you have heard the word "gallon" used with.

- a 5-gallon bucket
- a 1-gallon carton of milk
- a 1/2-gallon carton of milk
- a car's gas tank is so many gallons
- a water heater can hold so many gallons
- a bathtub can hold so many gallons
- A very large pot can hold 1 gallon of soup or stew.

- _____
- _____
- _____

One gallon of water is so much that you can fill FOUR quarts out of it.

   It would be a lot to drink!

11. Test yourself!

    Put one gallon of water into a bucket, preferably a 5-gallon bucket. (Four quarts make one gallon.) Can you carry it?

    Put another gallon of water into the same bucket. Can you still carry it?

    Now put 1 gallon of water into another bucket as well, and try to carry one such bucket in each hand. Can you?

    How many gallons of water can you carry using two buckets?

    Water can get pretty heavy!

How many quarts of water fit into a 5-gallon bucket?

Puzzle Corner

Hint: Think first how many quarts of water fit into 1 gallon, 2 gallons, and so on.

Milliliters and Liters

This is a measuring cup that measures volume in **milliliters (ml)**. Milliliters are very tiny units—you need lots of them to measure, for example, the volume of a glass.

This measuring cup goes up to 500 ml. And, 500 ml is exactly 1/2 liter. You can see that written near the top of the measuring cup.

A **liter** is 1,000 milliliters. A liter is abbreviated **l**, or sometimes with a capital L.

One liter is <u>very close to a quart</u> (just a little bit more).

1. Measure the volume of a few cups, glasses, jars, and other small containers. You will need a measuring cup that measures in milliliters.

Item	Volume in milliliters

2. Measure 1 liter of water into a pan. Then *guess* how many liters of water will fit into your pan.

    My Guess: the pan will hold _____ liters of water.

    Now, measure another liter of water into the pan, and another, until it is full. In the end, you can pour in 100 ml of water at a time.

    The pan holds _____ L _____ ml of water.

3. Measure the volume of another pan using the same method. First *guess* how many liters of water will fit into your pan.

    My Guess: the pan will hold _____ liters of water.

    Measurement: the pan holds _____ L _____ ml of water.

4. At home, find food products or personal care products that show the volume on the label using milliliters and/or liters.

Item	Volume

This section is optional.

5. Remember that 1 liter is 1,000 milliliters. Convert between liters and milliliters.

a.	b.	c.
1 L = _____ ml	1 L 200 ml = _____ ml	7 L 70 ml = _____ ml
2 L = _____ ml	5 L 490 ml = _____ ml	4 L 3 ml = _____ ml
6 L = _____ ml	4 L 230 ml = _____ ml	9 L 409 ml = _____ ml

6. One shampoo bottle contains 1 liter of shampoo. Another one contains 478 ml. How much more does the bigger one contain?
Hint: Change the 1 liter into milliliters.

7. How much liquid is in three water bottles that contain 450 ml each? Give your answer in liters and milliliters.

8. How many 250-ml glasses can you fill from a 1-liter bottle of juice?

   And how many 200-ml glasses?

9. Out of a 2-liter pitcher full of juice, Mom poured 5 glasses of 250 ml each. How much liquid is left in the pitcher?

Mixed Review Chapter 8

1. Estimate the answers by rounding the numbers to the nearest ten or nearest hundred. Then find the exact answers.

   a. A desk costs $154 and chairs cost $128. First, estimate the total bill. Then find the exact total.

   My estimate: about $_____

   b. Ed bought two computers for $1,298 each. First, estimate the total bill. Then find the exact total.

   My estimate: about $_____

   c. One TV costs $1,255 and another costs $787. Estimate the price difference between the two. Then find the exact difference.

   My estimate: about $_____

2. A rectangular pathway is 90 feet long and 6 feet wide. What is its area?

3. A large room is 20 feet by 9 feet. It is then divided into two equal parts. What is the area of one part?

4. Write the numbers immediately before and after the given number.

   a. _____, 2,778, _____   **b.** _____, 6,060, _____

   c. _____, 7,150, _____   **d.** _____, 7,000, _____

115

5. Multiply.

a. $5 \times 6 = $ ___	b. $6 \times 7 = $ ___	c. $9 \times 9 = $ ___
$3 \times 6 = $ ___	$4 \times 7 = $ ___	$8 \times 8 = $ ___
$8 \times 9 = $ ___	$5 \times 12 = $ ___	$6 \times 9 = $ ___
$7 \times 7 = $ ___	$8 \times 12 = $ ___	$6 \times 12 = $ ___

6. Break each multiplication into another where you multiply three numbers, one of them being 10. Multiply and fill in.

a. 7×30 $= $ ___ $\times$ ___ $\times 10$ $= $ ___ $\times 10 = $ ___	b. 5×60 $= $ ___ $\times$ ___ $\times 10$ $= $ ___ $\times 10 = $ ___

7. Multiply using the shortcut.

a. $8 \times 70 = $ ___	b. $3 \times 80 = $ ___	c. $50 \times 4 = $ ___
d. $30 \times 9 = $ ___	e. $20 \times 6 = $ ___	f. $4 \times 90 = $ ___

8. Find the total area of this rectangle, and also the area of each little part.

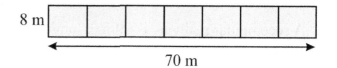

8 m, 70 m

Total area = _____

Area of each part = _____

9. Subtract. Check by adding.

a.   7 2 6 2
    − 2 3 1 6   + _____
	b. 6 0 0 3
	    − 3 2 4 2 + _____

Review Chapter 8

1. Draw lines of these lengths:

    a. 4 1/4 in

    b. 5 cm 7 mm

2. Measure the sides of this triangle in centimeters and millimeters, and find its perimeter.

    AB: _____ cm _____ mm

    BC: _____ cm _____ mm

    CA: _____ cm _____ mm

    perimeter: _____ cm _____ mm

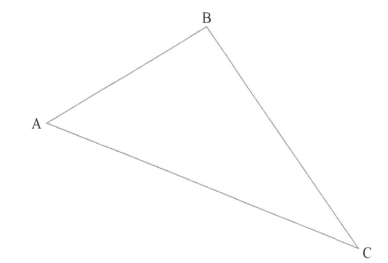

3. Measure the sides of this quadrilateral to the nearest quarter inch, and find its perimeter.

    AB: _____ in.   BC: _____ in.

    CD: _____ in.   DA: _____ in.

    perimeter: _____ in.

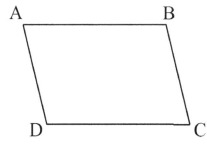

4. Write or say in order from the smallest to the biggest unit: cm km m mm

5. Write or say in order from the smallest to the biggest unit: ft in yd mi

6. Write or say in order from the smallest to the biggest unit: gal pt C qt

7. Name two different units you can use to measure the weight of people.

8. Fill in the blanks with suitable units of length. Sometimes several different units are possible

   a. A butterfly's wings were 6 _____ wide.   **b.** Sherry is 66 _____ tall.

   c. Jessica jogged 5 _____ yesterday.   **d.** The box was 60 _____ tall.

   e. The distance from the city   **f.** The room was 4 _____ wide.
   to the little town is 80 _____ .

   g. The eraser is 3 _____ long

9. Write the weight the scales are showing.

   a. _____ lb _____ oz   **b.** _____ lb _____ oz   **c.** _____ lb _____ oz

10. Have your teacher give you a small object. Use the scale to find out how much it weighs in either pounds and ounces, or in grams.

    It weighs _____ .

11. Have your teacher give you a small container. Use a measuring cup to find out how much water it can hold in milliliters.

    It holds _____ ml.

12. Fill in the blanks with suitable units of weight and volume. Sometimes several different units are possible

    a. Mom bought 5 _____ of apples.   **b.** Mary drank 350 _____ of juice.

    c. Dr. Smith weighs about 70 _____ .   **d.** The banana weighed 3 _____ .

    e. The pan holds 2 _____ of water.   **f.** A cell phone weighs about 100 _____ .

Chapter 9: Division
Introduction

The ninth chapter of *Math Mammoth Grade 3* covers the concept of division, basic division facts that are based on the multiplication tables, and the concept of remainder. The aim is to lay a good foundation for the concept of division, cementing the link between multiplication and division.

The concept of division in itself is not difficult—after all, it is like backwards multiplication. From that follows that the student needs to know the multiplication tables well as a prerequisite for this chapter. The student can start studying the lessons in this chapter even if he still needs some practice with the multiplication tables, but if he is a long ways from mastering them, he should not study this chapter yet.

There are basically two ways to illustrate division with concrete objects. The first way is equal sharing: we divide or share items equally between people. For example, the problem 12 ÷ 3 would mean, "If you share 12 bananas equally between 3 people, how many bananas does each one get?"

The second way has to do with grouping. The problem 12 ÷ 3 would be, "If you have 12 items, how many groups of three items can you make?" These two interpretations of division are important to understand so that the student can solve real-life and mathematical problems involving division.

We also study division by zero. From studying that lesson, students should recognize that division by zero "does not work." I realize that in higher forms of mathematics, division by zero may be defined (such as 1 ÷ 0 = infinity), but for now, this is the understanding that a third grader should get.

Lastly, students study the concept of remainder, or division that is not exact. We start by letting the students find the remainder using visual models (you could also use manipulatives). Then they learn how to find the remainder by calculating. This concept will be studied again in fourth grade.

The Lessons

	page	span
Division as Making Groups	122	*4 pages*
Division and Multiplication	126	*4 pages*
Division and Multiplication Facts	130	*3 pages*
Dividing Evenly into Groups	133	*4 pages*
Division Word Problems	137	*3 pages*
Zero in Division	140	*3 pages*
When Division Is Not Exact	143	*3 pages*
More Practice with the Remainder	146	*2 pages*
Mixed Review Chapter 9	148	*2 pages*
Review Chapter 9	150	*2 pages*

Helpful Resources on the Internet

Use these online resources as you see fit to supplement the main text.

CONCEPT OF DIVISION

Sharing
Solve word problems involving sharing. Choose "with remainders".
http://www.topmarks.co.uk/Flash.aspx?f=sharingv2

Airline Grouping
Practice division as making groups.
http://www.ictgames.com/airlineGrouping/airlineGrouping.html

Division by Sharing Fish
Drag and drop the fish into the fish tanks to "share" them. Fill in the empty boxes.
http://www.snappymaths.com/multdiv/earlymultdiv/interactive/sharing2/sharingframe.htm

DIVISION FACTS

Puzzle Pics Division
Drag the puzzle pieces to the correct answers and reveal the mystery picture!
http://www.mathplayground.com/puzzle_pics_division.html

Flying High Division
Fly your plane safely through the storm clouds by answering questions correctly.
http://www.multiplication.com/games/play/flying-high-division

Math Mahjong - Division
Try to match all the tiles.
http://www.sheppardsoftware.com/mathgames/mahjong/mahjongMath_division_easy.htm

Bike Racing Math Division
Win the race by clicking on the correct answer to speed up the motorcycle.
http://www.mathnook.com/math/bike-racing-math-division.html

Operation Snowman
Choose which operation you would use to solve the word problems.
http://www.harcourtschool.com/activity/operation_snowman/

Division Flashing Numbers
Divide each number by the given number and click on the flashing sign beneath when it is showing the right answer.
http://www.topmarks.co.uk/Flash.aspx?b=maths/division

Math Magician games
Practice division skills with these interactive online flashcards. Answer 20 questions in one minute.
http://web.archive.org/web/20160828220841/http://oswego.org/ocsd-web/games/mathmagician/mathsdiv.html

Cross the Swamp
Help Little Ron move from log to log across the swamp and practice multiplication/division or addition/subtraction.
http://www.bbc.co.uk/schools/starship/maths/crosstheswamp.shtml

Tux Math
A free software for practicing math facts with many options. Includes all operations. You need to shoot falling comets that can damage penguins' igloos.
http://sourceforge.net/projects/tuxmath

MISCELLANEOUS

Dividing by Zero at Math Is Fun
This page gives illustrations of why division by zero is undefined.
https://www.mathsisfun.com/numbers/dividing-by-zero.html

Division Facts with Remainders
Type the answers into the boxes and click "check".
http://www.mathplayground.com/division02.html

Rags to Riches Word Problems
Solve math problems about Water Park and you will win tickets to the park!
https://www.quia.com/rr/10249.html

Word Problems with Katie
Practice multiplication and division with these simple word problems.
http://www.mathplayground.com/WordProblemsWithKatie2.html

Division as Making Groups

There are 12 daisies. Make groups of 3.

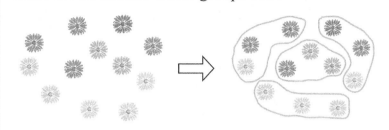

How many groups? *Four groups.*

How many 3's are there in 12? *Four.*

1. Divide into groups.

a. There are __15__ carrots. Make groups of 5.  How many groups? _____ How many 5's are there in __15__? _____	**b.** There are _____ berries. Make groups of 4.  How many groups? _____ How many 4's are there in _____? _____	**c.** There are _____ apples. Make groups of 3.  How many groups? _____ How many 3's are there in _____? _____
d. There are _____ fish. Make groups of 2.  How many groups? _____ How many 2's are there in _____? _____	**e.** There are _____ daisies. Make groups of 6.  How many groups? _____ How many 6's are there in _____? _____	**f.** There are _____ camels. Make groups of 4. How many groups? _____ How many 4's are there in _____? _____

DIVIDE... 12 dogs into groups of **four**.

How many groups? _Three_

How many 4's in 12? _____

$12 \div 4 = 3$

"Twelve divided by four is three."

DIVIDE... 15 elephants into groups of **three**.

How many groups? _____

How many 3's in 15? _____

$15 \div 3 = 5$

"Fifteen divided by three is five."

$\underline{18 \div 6 = ?}$ *Think*: If you DIVIDE 18 into groups of six, how many groups are there? How many groups of six are there in 18? How many sixes are there in 18?

Since $6 + 6 + 6 = 18$, there are THREE sixes in 18. So, $18 \div 6 = 3$

2. Write a division sentence to fit the pictures in exercise 1.

a. _____ ÷ _____ = _____   b. _____ ÷ _____ = _____   c. _____ ÷ _____ = _____

d. _____ ÷ _____ = _____   e. _____ ÷ _____ = _____   f. _____ ÷ _____ = _____

3. Make a division sentence.

a. Divide 10 rams into groups of two. How many groups?

_____ ÷ _____ = _____

b. Divide _____ camels into groups of four. How many groups?

_____ ÷ _____ = _____

c. Divide _____ apples into groups of six. How many groups?

_____ ÷ _____ = _____

d. Divide _____ books into groups of three. How many groups?

_____ ÷ _____ = _____

e. Divide _____ scissors into groups of five. How many groups?

_____ ÷ _____ = _____

f. Divide _____ crosses into groups of three. How many groups?

✲✲✲✲✲✲✲
✲✲✲✲✲✲
✲✲✲✲✲✲

_____ ÷ _____ = _____

4. Draw sticks. Divide them into groups to fit the division sentence.

a. 18 ÷ 3 = _____	**b.** 24 ÷ 2 = _____
c. 21 ÷ 3 = _____	**d.** 25 ÷ 5 = _____
e. 15 ÷ 5 = _____	**f.** 24 ÷ 8 = _____

5. Make groups by circling dots and write a division sentence.

a. Make groups of 4	**b.** Make groups of 2	**c.** Make groups of 6	**d.** Make groups of 3
_____ ÷ 4 = _____	_____ ÷ 2 = _____	_____ ÷ 6 = _____	_____ ÷ 3 = _____
e. Make groups of 5	**f.** Make groups of 7	**g.** Make groups of 6	**h.** Make groups of 10
_____ ÷ 5 = _____	_____ ÷ 7 = _____	_____ ÷ 6 = _____	_____ ÷ 10 = _____

6. Solve the word problems. Write a division or a multiplication for each problem. The box □ is for the × or ÷ symbol.

a. The class has 20 students. You can fit five students into a van. How many vans are needed? ____ □ ____ = ____	**b.** Ken placed 30 marbles in rows of 5. How many rows did he get? ____ □ ____ = ____
c. Erica packed hairpins in bags. She put 20 pins in each bag and filled four bags. How many pins were there? ____ □ ____ = ____	**d.** Kelly packaged 28 T-shirts in bags. She put four shirts in each bag. How many bags did she use? ____ □ ____ = ____
e. Brian has 16 poster boards. He needs four of them to make a big poster board. How many big ones can he make? ____ □ ____ = ____	**f.** Marlene studied three hours each day for seven days. How many hours did she spend studying in total? ____ □ ____ = ____

7. Solve. You can draw to help. Can you find a pattern?

a.	b.	c.
4 ÷ 2 = ____	20 ÷ 10 = ____	10 ÷ 5 = ____
6 ÷ 2 = ____	30 ÷ 10 = ____	15 ÷ 5 = ____
8 ÷ 2 = ____	40 ÷ 10 = ____	20 ÷ 5 = ____
10 ÷ 2 = ____	50 ÷ 10 = ____	25 ÷ 5 = ____
12 ÷ 2 = ____	____ ÷ 10 = ____	____ ÷ 5 = ____
14 ÷ 2 = ____	____ ÷ 10 = ____	____ ÷ 5 = ____
16 ÷ 2 = ____	____ ÷ 10 = ____	____ ÷ 5 = ____
____ ÷ 2 = ____	____ ÷ 10 = ____	____ ÷ 5 = ____
____ ÷ 2 = ____	____ ÷ 10 = ____	____ ÷ 5 = ____

Division and Multiplication

We get both a **multiplication fact** and a **division fact** from the same picture:	Three **groups of 4** makes 12. 3 × 4 = 12 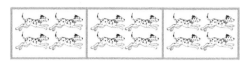 12 divided into **groups of 4** is three groups. 12 ÷ 4 = 3

Multiplication and division are very closely related. They are opposite operations. You could say division is "backwards" multiplication.

1. Fill in the blanks.

a. Two **groups of 6** is 12. 2 × 6 = 12  12 divided into **groups of 6** is two groups. 12 ÷ 6 = 2	**b.** Five **groups of 2** is _____. ____ × 2 = ____  ____ divided into **groups of 2** is ____ groups. _____ ÷ 2 = ____
c. One **group of 4** is 4. ____ × 4 = ____ 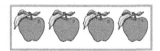 4 divided into **groups of 4** is one group. _____ ÷ 4 = ____	**d.** ____ **groups of 3** is _____. ____ × ____ = ____  ____ divided into **groups of 3** is ____ groups. _____ ÷ ____ = ____
e. Five **groups of 1** is 5. ____ × 1 = ____  5 divided into **groups of 1** is ____ groups. _____ ÷ 1 = ____	**f.** ____ **groups of** ____ is _____. ____ × ____ = ____ 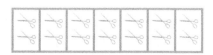 ____ divided into **groups of 2** is ____ groups. _____ ÷ ____ = ____

2. Make groups. Then write the division and multiplication facts that the pictures illustrate.

a. Make groups of four. ___ × 4 = 8 8 ÷ 4 = ___	**b.** Make groups of two. ___ × 2 = ___ ___ ÷ 2 = ___
c. Make groups of four. ___ × 4 = ___ ___ ÷ 4 = ___	**d.** Make groups of six. ___ × 6 = ___ ___ ÷ 6 = ___
e. ___ × 4 = ___ ___ ÷ 4 = ___	**f.** ___ × 7 = ___ ___ ÷ 7 = ___

g. ___ × 6 = ___ ___ ÷ 6 = ___	**h.** ___ × 2 = ___ ___ ÷ 2 = ___	**i.** ___ × 5 = ___ ___ ÷ 5 = ___

3. Now draw sticks or circles and make a picture yourself. Write the division and multiplication sentences.

a. Draw 15 sticks. Make groups of 5. ___ × 5 = ___ ___ ÷ 5 = ___	**b.** Draw 24 sticks. Make groups of 8. ___ × ___ = ___ ___ ÷ ___ = ___	**c.** Draw 30 sticks. Make groups of 5. ___ × ___ = ___ ___ ÷ ___ = ___

d. Draw 27 sticks. Make groups of 9.	**e.** Draw 32 sticks. Make groups of 16.	**f.** Draw 16 sticks. Make groups of 2.
___ × ___ = ___ ___ ÷ ___ = ___	___ × ___ = ___ ___ ÷ ___ = ___	___ × ___ = ___ ___ ÷ ___ = ___
g. Draw 8 sticks. Make a group of 8.	**h.** Draw 18 sticks. Make groups of 9.	**i.** Draw 20 sticks. Make groups of 5.
___ × ___ = ___ ___ ÷ ___ = ___	___ × ___ = ___ ___ ÷ ___ = ___	___ × ___ = ___ ___ ÷ ___ = ___

4. For each multiplication fact, also write a division fact. Think about the groups!

a. $7 \times 2 = $ ___ ___ $\div 2 = $ ___	**b.** $12 \times 2 = $ ___ ___ $\div 2 = $ ___	**c.** $8 \times 5 = $ ___ ___ $\div 5 = $ ___
d. $6 \times 7 = $ ___ ___ ÷ ___ = ___	**e.** $7 \times 7 = $ ___ ___ ÷ ___ = ___	**f.** $11 \times 3 = $ ___ ___ ÷ ___ = ___
g. $9 \times 8 = $ ___ ___ ÷ ___ = ___	**h.** $1 \times 5 = $ ___ ___ ÷ ___ = ___	**i.** $7 \times 9 = $ ___ ___ ÷ ___ = ___

You can solve a division problem by thinking of the *matching multiplication*.

$30 \div 6 =$ ___

___ $\times 6 = 30$

Think: what times 6 is 30?

So, since you already know the multiplication tables, division will be easy!

5. For each division, think of the corresponding multiplication and solve.

a. $14 \div 2 =$ _____ ___ $\times 2 = 14$	b. $18 \div 2 =$ _____ ___ $\times 2 =$ _____	c. $21 \div 7 =$ _____ ___ $\times 7 =$ _____
d. $54 \div 6 =$ _____ ___ $\times$ ___ $=$ ___	e. $24 \div 4 =$ _____ ___ $\times$ ___ $=$ ___	f. $30 \div 3 =$ _____ ___ $\times$ ___ $=$ ___
g. $32 \div 4 =$ _____ ___ $\times$ ___ $=$ ___	h. $56 \div 7 =$ _____ ___ $\times$ ___ $=$ ___	i. $55 \div 5 =$ _____ ___ $\times$ ___ $=$ ___

6. Divide. Again **think** of the multiplication.

a.	b.	c.	d.
$24 \div 4 =$ ___	$15 \div 5 =$ ___	$32 \div 8 =$ ___	$48 \div 6 =$ ___
$16 \div 2 =$ ___	$35 \div 5 =$ ___	$40 \div 8 =$ ___	$56 \div 8 =$ ___
$20 \div 2 =$ ___	$49 \div 7 =$ ___	$50 \div 5 =$ ___	$81 \div 9 =$ ___
$36 \div 9 =$ ___	$54 \div 9 =$ ___	$42 \div 6 =$ ___	$100 \div 10 =$ ___

Puzzle Corner Think of multiplication, and solve.

a. $1{,}000 \div 100 =$ _____   b. $400 \div 50 =$ _____   c. $200 \div 4 =$ _____

d. $1{,}000 \div 500 =$ _____   e. $800 \div 800 =$ _____   f. $200 \div 40 =$ _____

Division and Multiplication Facts

From the same picture, you can actually get
two multiplication facts AND **two** division facts:

Bananas divided into rows:	The same bananas divided into columns:

 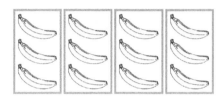

12 bananas in groups of four is three groups. $12 \div 4 = 3$ $3 \times 4 = 12$

12 bananas in groups of three is four groups. $12 \div 3 = 4$ $4 \times 3 = 12$

Just like with addition and subtraction, we can form **fact families** that have two multiplication facts and two division facts.

1. Make two division sentences and two multiplication sentences out of the same picture.

a.

$4 \times 6 =$ ____

$6 \times 4 =$ ____

____ $\div 4 =$ ____

____ $\div 6 =$ ____

b.

____ $\times$ ____ $=$ ____

____ $\times$ ____ $=$ ____

____ $\div$ ____ $=$ ____

____ $\div$ ____ $=$ ____

c.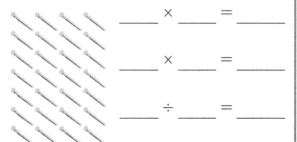

____ $\times$ ____ $=$ ____

____ $\times$ ____ $=$ ____

____ $\div$ ____ $=$ ____

____ $\div$ ____ $=$ ____

d.

____ $\times$ ____ $=$ ____

____ $\times$ ____ $=$ ____

____ $\div$ ____ $=$ ____

____ $\div$ ____ $=$ ____

2. Make two division sentences and two multiplication sentences out of the same picture.

a.	b.
____ × ____ = ____	____ × ____ = ____
____ × ____ = ____	____ × ____ = ____
____ ÷ ____ = ____	____ ÷ ____ = ____
____ ÷ ____ = ____	____ ÷ ____ = ____
c.	d.
____ × ____ = ____	____ × ____ = ____
____ × ____ = ____	____ × ____ = ____
____ ÷ ____ = ____	____ ÷ ____ = ____
____ ÷ ____ = ____	____ ÷ ____ = ____

3. Here, each division has a matching multiplication. Fill in the missing parts.

a. $12 \div 2 =$ ____	b. ____ $\div 2 =$ ____	c. $16 \div$ ____ $=$ ____
____ $\times 2 = 12$	____ $\times 2 = 22$	____ $\times 2 = 16$
d. ____ $\div$ ____ $= 8$	e. $32 \div 4 =$ ____	f. ____ $\div 5 = 5$
$8 \times 3 =$ ____	____ $\times 4 =$ ____	____ $\times 5 =$ ____

4. Divide. Think of the corresponding multiplication problem!

a.	b.	c.	d.
$18 \div 2 =$ ____	$15 \div 3 =$ ____	$40 \div 4 =$ ____	$45 \div 5 =$ ____
$16 \div 2 =$ ____	$18 \div 3 =$ ____	$16 \div 4 =$ ____	$55 \div 5 =$ ____
$24 \div 2 =$ ____	$21 \div 3 =$ ____	$36 \div 4 =$ ____	$60 \div 5 =$ ____

5. Find the missing numbers in these divisions. Check by writing a matching multiplication.

a. _____ ÷ 2 = 7 ___ × ___ = _____	b. _____ ÷ 5 = 6 ___ × ___ = _____	c. _____ ÷ 7 = 4 ___ × ___ = _____
d. 20 ÷ ___ = 2 ___ × ___ = _____	e. 35 ÷ ___ = 5 ___ × ___ = _____	f. 56 ÷ ___ = 8 ___ × ___ = _____

6. Write a multiplication *and* a division for each situation.

a. Sandra put 5 toys in each of the ten boxes.	___ × ___ = _____ _____ ÷ ___ = ___
b. Twenty toys were put into boxes so that there were four toys in each box.	___ × ___ = _____ _____ ÷ ___ = ___
c. Ken read three little books in one afternoon. Each book had 20 pages.	___ × ___ = _____ _____ ÷ ___ = ___
d. Three children shared the job of planting 30 apple trees.	___ × ___ = _____ _____ ÷ ___ = ___

7. Find the missing numbers. Think of the corresponding multiplication problem!

a.	b.	c.	d.
56 ÷ 7 = ___	48 ÷ 6 = ___	___ ÷ 9 = 5	48 ÷ ___ = 6
70 ÷ 7 = ___	24 ÷ 6 = ___	___ ÷ 9 = 3	24 ÷ ___ = 2
42 ÷ 7 = ___	66 ÷ 6 = ___	___ ÷ 9 = 2	72 ÷ ___ = 8
49 ÷ 7 = ___	72 ÷ 6 = ___	___ ÷ 9 = 9	40 ÷ ___ = 4
28 ÷ 7 = ___	54 ÷ 6 = ___	___ ÷ 9 = 11	32 ÷ ___ = 8

Dividing Evenly into Groups

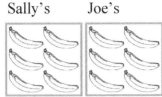

If we divide 12 bananas evenly between Joe and Sally, how many does each one get?

Both Joe and Sally each get 6 bananas.

We can write the DIVISION 12 ÷ 2 = 6.

When things are divided or shared equally, we can write a division.

1. Two children are sharing. Divide the things into **two** equal groups. Write a division.

| a.  ___ ÷ 2 = ___ Each child gets ___ . | b. 🍎🍎🍎🍎🍎🍎 ___ ÷ 2 = ___ Each child gets ___ . | c. 🍎🍎🍎🍎🍎 🍎🍎🍎🍎🍎 ___ ÷ ___ = ___ Each child gets ___ . |

2. Three children are sharing. Divide the things into **three** equal groups. Write a division.

| a. (12 cubes) ___ ÷ 3 = ___ Each child gets ___ . | b. (6 apples) ___ ÷ 3 = ___ Each child gets ___ . | c. (grid) ___ ÷ ___ = ___ Each child gets ___ . |

3. Four children are sharing. Divide the things into **four** equal groups. Write a division.

| a. (8 apples) ___ ÷ 4 = ___ | b. (12 pineapples) 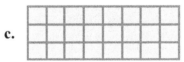 ___ ÷ 4 = ___ | c. (24 apples) ___ ÷ ___ = ___ |

Let's think about the division 18 ÷ 3 in **TWO** different ways.

1) We have 18 carrots, and we will make **_groups of 3_**. How many groups do we get?

Six groups. So, 18 ÷ 3 = 6.

2) We divide the 18 carrots evenly **_into three groups_**, like sharing them among three people. How many are there in each group?

Six. So, 18 ÷ 3 = 6.

There are TWO ways to think about division:

1) You make groups of a certain size. How many groups do you get?

2) You make a certain number of groups, dividing the things equally into these groups. How many are there in each group?

4. Divide things evenly into groups.

a. Divide into two groups. 8 ÷ 2 = ____	**b.** Divide into five groups. ____ ÷ ____ = ____
c. Divide into one group. ____ ÷ ____ = ____	**d.** Divide into four groups. ____ ÷ ____ = ____

e. Make 3 groups 21 ÷ 3 = ____	f. Make 1 group ____ ÷ 1 = ____	g. Make 10 groups ____ ÷ 10 = ____	h. Make 2 groups ____ ÷ 2 = ____

5. Divide. Remember to think about the multiplication problem.

a. 40 ÷ 8 = ____ 6 ÷ 3 = ____ 16 ÷ 2 = ____	**b.** 48 ÷ 12 = ____ 60 ÷ 6 = ____ 25 ÷ 5 = ____	**c.** 36 ÷ 9 = ____ 36 ÷ 6 = ____ 56 ÷ 7 = ____
d. 30 ÷ 5 = ____ 24 ÷ 3 = ____ 64 ÷ 8 = ____	**e.** 99 ÷ 9 = ____ 72 ÷ 6 = ____ 27 ÷ 3 = ____	**f.** 100 ÷ 10 = ____ 80 ÷ 10 = ____ 45 ÷ 9 = ____

6. Find the unknown numbers (marked by a circle or ?).

a. 16 ÷ 4 = ? ? = ____	**b.** 21 ÷ ? = 3 ? = ____	**c.** 42 ÷ ? = 6 ? = ____	**d.** ? ÷ 5 = 12 ? = ____
e. ◯ ÷ 4 = 7 ◯ = ____	**f.** 54 ÷ ◯ = 6 ◯ = ____	**g.** 144 ÷ 12 = ◯ ◯ = ____	**h.** ◯ ÷ 11 = 11 ◯ = ____

7. Solve. Write a division or a multiplication for each problem. The box ☐ is where you will write either × or ÷.

a. Amanda, Jill, and Bill shared evenly 18 marbles in a game. How many marbles did each one get? ____ ☐ ____ = ____	**b.** Four children played marbles. Each one had 7 marbles. How many marbles were there in total? ____ ☐ ____ = ____
c. Ashley cut a 24-inch long string into 6 equal pieces. How long was each piece of string? ____ ☐ ____ = ____	**d.** Mom bought 24 hairpins and divided them evenly among her 3 daughters. How many hairpins did each girl get? ____ ☐ ____ = ____

8. **a.** Write or make a *division* story problem about 20 apples and some horses.

b. Write or make a *division* story problem about 24 toy cars and some children.

9. Fill in the division tables!

a. Division table of six	b. Division table of seven	c. Division table of eight
6 ÷ 6 = ___	7 ÷ 7 = ___	8 ÷ 8 = ___
12 ÷ 6 = ___	14 ÷ 7 = ___	16 ÷ 8 = ___
___ ÷ 6 = ___	___ ÷ 7 = ___	___ ÷ 8 = ___
___ ÷ 6 = ___	___ ÷ 7 = ___	___ ÷ 8 = ___
___ ÷ 6 = ___	___ ÷ 7 = ___	___ ÷ 8 = ___
___ ÷ 6 = ___	___ ÷ 7 = ___	___ ÷ 8 = ___
___ ÷ 6 = ___	___ ÷ 7 = ___	___ ÷ 8 = ___
___ ÷ 6 = ___	___ ÷ 7 = ___	___ ÷ 8 = ___
___ ÷ 6 = ___	___ ÷ 7 = ___	___ ÷ 8 = ___
___ ÷ 6 = ___	___ ÷ 7 = ___	___ ÷ 8 = ___
___ ÷ 6 = ___	___ ÷ 7 = ___	___ ÷ 8 = ___
___ ÷ 6 = ___	___ ÷ 7 = ___	___ ÷ 8 = ___

Notice the patterns in these tables! How are they similar to the multiplication tables?

Division Word Problems

With both **multiplication** and **division** you have groups that are the same size.

Six *groups of 3* makes a total of 18. $6 \times 3 = 18$

18 divided into *groups of 3* is six groups. $18 \div 3 = 6$

In multiplication word problems

- There are groups that are the same size.
- You are asked the total.
- You know *how many groups there are* and *how many are in each group*.

In division word problems

- There are groups that are the same size.
- You already know the total.
- You are asked *how many groups there are* or *how many are in each group*.

1. Solve. Write a division or a multiplication for each problem.
   Think: is the problem asking for a total? Or do you already know the total, and it asks "how many groups/parts" or "how many in each group/part"?
   The box ☐ is where you will write either × or ÷ .

a. Henry has 90 stamps in his stamp book with ten stamps on each page. How many pages are full of stamps? ___ ☐ ___ = ___	**b.** Jill put twelve stamps per page in her stamp book. She has eight pages full of stamps. How many stamps does she have? ___ ☐ ___ = ___
c. If four children can fit into one taxi, how many children would there be in 11 taxis? ___ ☐ ___ = ___	**d.** Four children can fit into one taxi. How many taxis do you need for 12 children? ___ ☐ ___ = ___

2. Solve. Write a division or a multiplication for each problem. Think: is the problem asking for a total? Or do you already know the total, and it asks "how many groups/parts" or "how many in each group/part"?

a. If there are ten eggs in each carton, how many eggs are in five cartons?

____ ☐ ____ = ____

b. Jack placed ten toy cars in bags with five cars in each bag. How many bags did he use?

____ ☐ ____ = ____

c. Cindy can fit three bottles of juice into one plastic bag. How many can she fit into five bags?

____ ☐ ____ = ____

d. Cindy can fit three bottles of juice into one plastic bag. How many bags will she need for 18 bottles?

____ ☐ ____ = ____

e. Sally, Joe, and Tammy divided 36 cherries equally. How many did each one get?

____ ☐ ____ = ____

f. The teacher wanted to make five groups out of a class of 25 students. How many students were in each group?

____ ☐ ____ = ____

g. How many people are in seven vans if each van has five people in it?

____ ☐ ____ = ____

h. Joe divided a 27-inch long board into three parts. How long was each part?

____ ☐ ____ = ____

i. Mom made 20 liters of tea. She put it into 2-liter containers. How many containers did she fill?

____ ☐ ____ = ____

j. Harry jogged a 9-mile long track in three equal segments, resting in between. How long was each part?

____ ☐ ____ = ____

k. Mom has 24 eggs. It takes eight eggs to make an omelet for the family. How many omelets can she make?

____ ☐ ____ = ____

l. If you can fit 12 crayons into a box, how many boxes do you need for 60 crayons?

____ ☐ ____ = ____

3. Solve. Write a number sentence(s) for each.

a. There were seven rows, with ten chairs in each row, PLUS one more row with eight chairs. How many chairs were there?	_____ ☐ _____ ☐ _____ = _____
b. Mom had 14 cherries in one container and 13 in another. Mom, Dad, and Jane shared them equally. How many did each get?	
c. There are four buckets of blueberries. They weigh 4 kg, 6 kg, 7 kg, and 5 kg. Mom packaged the blueberries into containers, 2 kg in each. How many containers did she use?	
d. One box holds 12 crayons. How many crayons are in four full boxes and one with five missing?	

4. Write the fact families.

a.
___ × ___ = ___
___ × ___ = ___
___ ÷ 9 = 4
___ ÷ ___ = ___

b.
___ × ___ = ___
___ × ___ = ___
___ ÷ ___ = ___
54 ÷ ___ = 6

c.
___ × 7 = 42
___ × ___ = ___
___ ÷ ___ = ___
___ ÷ ___ = ___

What numbers can go into the puzzles? The last one is totally empty so you can make one puzzle of your own!

Puzzle Corner

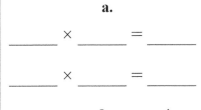

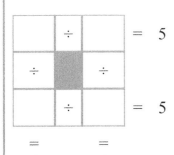

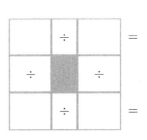

139

Zero in Division

What do you think 6 ÷ 0 would be?

We could think of sharing 6 bananas between 0 persons, but that does not make sense. We cannot even talk about how many each one gets, because there is no one around.

We could think of making groups of 0. How many groups would you get? Again, you would not get anywhere, you would never get those 6 bananas put into groups of 0.

You might think that maybe 6 ÷ 0 = 0 or that each person gets zero bananas. Check it with multiplication! You would get 0 × 0 = 6, which is not true! So, 6 ÷ 0 = 0 does not work either.

Dividing six by zero (6 ÷ 0) is "undefined." Basically, you cannot do it.

What about 0 ÷ 0? Couldn't we say 0 ÷ 0 = 0?

0 ÷ 0 is hard. The answer could be zero, but actually the answer could be *any* number :

Let's say that 0 ÷ 0 = 2.   Check by multiplying: 2 × 0 = 0; OK. So 2 would work.
Let's say that 0 ÷ 0 = 0.   Check by multiplying: 0 × 0 = 0; OK. So 0 would work.
Let's say that 0 ÷ 0 = 11. Check by multiplying: 11 × 0 = 0; OK. So 11 would work.

So, we cannot find just ONE answer. We say that the answer cannot be determined.

Dividing a number by zero does not work.

What about zero divided by something? That is perfectly fine.

0 ÷ 5 = 0   "If there are zero bananas and five people, each person gets 0 bananas."

1. Divide. CROSS OUT all the problems that are impossible. Think about sharing bananas.

a. 4 ÷ 1 = ___	b. 14 ÷ 14 = ___	c. 1 ÷ 1 = ___	d. 0 ÷ 5 = ___
4 ÷ 0 = ___	0 ÷ 0 = ___	7 ÷ 0 = ___	5 ÷ 5 = ___
e. 0 ÷ 1 = ___	f. 0 ÷ 14 = ___	g. 0 ÷ 3 = ___	h. 10 ÷ 10 = ___
0 ÷ 4 = ___	14 ÷ 0 = ___	0 ÷ 1 = ___	1 ÷ 1 = ___

In multiplication, zero works just fine!	
Multiplication means you have many groups of the same size. You can find the total by adding. Therefore:	
$5 \times 0 = 0+0+0+0+0 = 0$ (five groups of zero items)	$0 \times 3 = 0$ (zero groups of three items)

2. Multiply. Then for each multiplication, make a division sentence _if possible_. Some divisions are not possible!

a. $6 \times 1 = $ _____ ____ ÷ ____ = ____	**b.** $0 \times 8 = $ _____ ____ ÷ ____ = ____	**c.** $5 \times 7 = $ _____ ____ ÷ ____ = ____
d. $10 \times 11 = $ _____ ____ ÷ ____ = ____	**e.** $1 \times 1 = $ _____ ____ ÷ ____ = ____	**f.** $1 \times 8 = $ _____ ____ ÷ ____ = ____
g. $0 \times 0 = $ _____ ____ ÷ ____ = ____	**h.** $5 \times 9 = $ _____ ____ ÷ ____ = ____	**i.** $9 \times 0 = $ _____ ____ ÷ ____ = ____

3. Solve, and write a number sentence for each word problem.

a. Sally had 30 kg of rice. She put equal amounts into six bags. How much was in each bag?	**b.** There are six minivan taxis at the airport, and each can hold seven passengers. How many passengers can they take in total?
c. Greg bought three cartons of eggs, with 12 eggs in each carton. How many eggs did he get? 	**d.** The airplane had 56 passengers. Each minivan taxi can hold seven passengers. How many taxis are needed to take these passengers to a hotel? 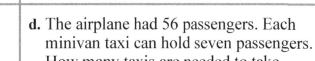 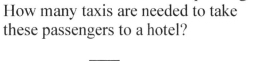
e. There are five tables, and each has four legs. How many legs are there in total? 	**f.** Kelly poured a total of four cups of milk into four glasses. How much milk was in each glass? 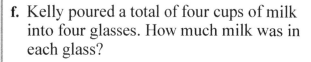

4. Divide. CROSS OUT all the problems that are impossible.

a. $9 \div 1 =$ ____ $9 \div 0 =$ ____	b. $0 \div 20 =$ ____ $20 \div 0 =$ ____	c. $11 \div 1 =$ ____ $8 \div 0 =$ ____	d. $0 \div 0 =$ ____ $0 \div 10 =$ ____

5. Make a QUESTION for each situation. (Think what you can find out using what the problem tells you.) Then solve your question.

a. Mark, Jack, and Joe decided to share their toy cars evenly in a game. Mark had 18 cars, Jack had 7, and Joe had 11.	**b.** Mrs. Elliott hired six children to do yard work. She paid one of them $15, and the rest of them $10 each.
c. The books on Alice's reading list have: 320, 129, 120, and 235 pages.	**d.** Jeremy wants to read two books that have 32 and 40 pages. He reads 12 pages a day.
	e. Kelly had 80 cm of red material and 40 cm of blue material. She cut it all into 20-cm pieces.
	f. A child arranged toy cars in rows of six cars. He made seven rows like that. The eighth row had three cars.

Puzzle Corner

Division and multiplication involving 0 leads to some funny or interesting situations. Can you solve what ☐ stands for?

a. $0 \div \square = 4$ b. $0 \times \square = 0$ c. $\square \div 0 = 6$ d. $0 \times \square = 3$

When Division is Not Exact

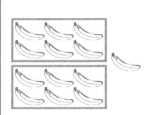

If you divide 13 bananas evenly between Joe and Sally, how many does each one get?

$13 \div 2 = ?$

Joe and Sally each get 6 bananas and one is left over. We write this as:

$13 \div 2 = 6, \; R1$

The leftover banana is called **the remainder**, and is indicated after the letter R.

(If we didn't want any leftovers, then both could get 6 ½ bananas.)

1. Fill in the blanks.

a. 14 bananas divided among 3 people gives 4 bananas to each and 2 bananas that cannot be divided evenly.

$14 \div 3 = 4,$ remainder 2

b. 14 carrots divided among 5 people gives 2 carrots to each and 4 carrots that cannot be divided evenly.

$14 \div 5 = 2,$ remainder 4

c. 8 scissors divided among 5 people gives 1 pair of scissors to each and 3 pairs that cannot be divided evenly.

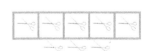

$8 \div 5 = \underline{},$ remainder ____

d. 3 apples divided among 5 people means we cannot share them equally. So, no one gets any apples. All 3 are left over.

$3 \div 5 = 0,$ remainder ____

e. ____ rams divided among 6 people gives ____ rams to each and ____ rams that cannot be divided evenly.

$\underline{} \div 6 = \underline{},$ remainder ____.

f. ____ camels divided between 2 people gives ____ camels to each person, and ____ camel left over.

$\underline{} \div 2 = \underline{},$ remainder ____.

Here's another way of looking at division and remainder. How many groups of 2 can we make out of 13 apples?

We can make six groups. One apple is left over.

13 ÷ 2 = 6 R1

2. Divide the dots into groups and write a division sentence.

a. Divide into groups of 3. 20 ÷ 3 = ____ remainder ____	**b.** Divide into groups of 4. 21 ÷ 4 = ____ remainder ____	**c.** Divide into groups of 6. ____ ÷ 6 = ____ remainder ____	**d.** Divide into groups of 5. ____ ÷ 5 = ____ remainder ____
e. Divide into groups of 7. ____ ÷ 7 = ____ remainder ____	**f.** Divide into groups of 9. ____ ÷ 9 = ____ remainder ____	**g.** Divide into groups of 3. ____ ÷ 3 = ____ remainder ____	**h.** Divide into groups of 5. ____ ÷ 5 = ____ remainder ____

4 ÷ 5 = ?

How many groups of 5 can we make out of 4 apples?

No groups. All four apples are left over.

4 ÷ 5 = 0 R4

3. Divide and indicate the remainders.

a. 7 ÷ 2 = ____, R ____ 1 ÷ 2 = ____, R ____	**b.** 3 ÷ 4 = ____, R ____ 11 ÷ 2 = ____, R ____	**c.** 18 ÷ 5 = ____, R ____ 7 ÷ 6 = ____, R ____

$20 \div 3 = ?$

Think: How many groups of 3 are there in 20?
Or: How many times does 3 fit into 20?

$6 \times 3 = 18$ and $7 \times 3 = 21$ (too much).

So 3 goes into 20 six times. Since $6 \times 3 = 18$ and 18 is 2 less than 20, the remainder is 2.

You will find the remainder by finding the <u>difference</u> between 20 and $6 \times 3 = 18$.

Example. $42 \div 8 = ?$

Think: How many times does 8 fit into 42? $5 \times 8 = 40$ and $6 \times 8 = 48$. So, 8 goes into 42 five times. And, 5×8 is 40. The remainder is the difference between 40 and 42, or 2. **So, $42 \div 8 = 5$ R2**

4. Practice some more!

a. $13 \div 5 = $ ___, R ___	b. $5 \div 8 = $ ___, R ___	c. $47 \div 6 = $ ___, R ___
$14 \div 5 = $ ___, R ___	$25 \div 8 = $ ___, R ___	$50 \div 6 = $ ___, R ___
d. $13 \div 2 = $ ___, R ___	e. $54 \div 8 = $ ___, R ___	f. $57 \div 7 = $ ___, R ___
$13 \div 5 = $ ___, R ___	$67 \div 8 = $ ___, R ___	$39 \div 9 = $ ___, R ___

5. Divide. What patterns do you notice?

a.	b.	c.
$21 \div 2 = $ ___, R ___	$21 \div 3 = $ ___, R ___	$21 \div 4 = $ ___, R ___
$22 \div 2 = $ ___, R ___	$22 \div 3 = $ ___, R ___	$22 \div 4 = $ ___, R ___
$23 \div 2 = $ ___, R ___	$23 \div 3 = $ ___, R ___	$23 \div 4 = $ ___, R ___
$24 \div 2 = $ ___, R ___	$24 \div 3 = $ ___, R ___	$24 \div 4 = $ ___, R ___
$25 \div 2 = $ ___, R ___	$25 \div 3 = $ ___, R ___	$25 \div 4 = $ ___, R ___
$26 \div 2 = $ ___, R ___	$26 \div 3 = $ ___, R ___	$26 \div 4 = $ ___, R ___
$27 \div 2 = $ ___, R ___	$27 \div 3 = $ ___, R ___	$27 \div 4 = $ ___, R ___
$28 \div 2 = $ ___, R ___	$28 \div 3 = $ ___, R ___	$28 \div 4 = $ ___, R ___
$29 \div 2 = $ ___, R ___	$29 \div 3 = $ ___, R ___	$29 \div 4 = $ ___, R ___
$30 \div 2 = $ ___, R ___	$30 \div 3 = $ ___, R ___	$30 \div 4 = $ ___, R ___

More Practice with the Remainder

Division can also be written this way. The answer goes on top of the line.	5 $9\overline{)4\,5}$ This is the same as $45 \div 9 = 5$.	$3\overline{)2\,1}$ This is the same as $21 \div 3$. Write the answer in the right place.

1. Divide.

a. $3\overline{)2\,4}$ b. $5\overline{)4\,0}$ c. $6\overline{)2\,4}$ d. $8\overline{)6\,4}$

e. $7\overline{)4\,9}$ f. $4\overline{)2\,8}$ g. $9\overline{)8\,1}$ h. $9\overline{)5\,4}$

You can also find the remainder **by subtracting**. Remember, it is the **difference**— the 'leftovers'.	7 $5\overline{)3\,6}$ How many times does 5 go into 36? Write the answer on top of the line.	7 $5\overline{)3\,6}$ $\underline{-3\,5}$ 1 Now multiply 7×5. Write 35 under 36. Subtract. You get 1. It is the remainder.

2. Divide and find the remainder by subtracting!

a. 7
    $3\overline{)2\,2}$
    $\underline{-2\,1}$

b. $5\overline{)2\,1}$ c. $3\overline{)1\,7}$ d. $8\overline{)2\,9}$

e. $7\overline{)2\,6}$ f. $6\overline{)5\,2}$ g. $9\overline{)3\,5}$ h. $4\overline{)3\,5}$

3. Divide and find the remainder by subtracting.

a. 3)28   b. 5)48   c. 6)58   d. 8)71

e. 7)61   f. 6)46   g. 9)75   h. 7)54

4. Write a number sentence for each word problem. Indicate a remainder if any.

a. Mom baked 29 rolls and divided them as evenly as she could among eight guests. How many rolls did each one get? Were any left over? ____ ☐ ____ = ____	b. Susan planted five rows of pepper plants. Each row had seven plants. How many plants did she plant in total? ____ ☐ ____ = ____
c. Each of the 12 students has six pencils, except Jim, who has three. What is the total number of pencils? _____	d. Pat wanted to have three rolls for each of the 12 guests. How many rolls did she need? ____ ☐ ____ = ____
e. Susan wants to organize 12 flower pots into nice even rows. Can she organize them into rows of three flower pots? Of four? Of five? Of six?	f. The town gardener wants to organize 36 flower plants evenly into rows. What kind of rows can he make? Rows of ____, rows of ____, rows of ____, rows of ____, rows of ____, rows of ____.
g. You gave 40 lollipops to 15 children. How many did each child get? Are there any left over? ____ ☐ ____ = ____	h. Tim packed 56 eggs into cartons of 12 eggs each. How many cartons were full? ____ ☐ ____ = ____

Mixed Review Chapter 9

1. Find the area and perimeter of this rectangle. Use a ruler to measure its sides.

    Area:

    Perimeter:

2. Write these numbers in order from smallest to greatest.

    2513   5096   5606   5060   2466   2506   2516

    _____ < _____ < _____ < _____ < _____ < _____ < _____

3. Fill in.

    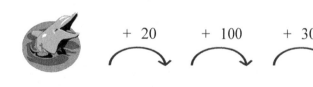 

    6880   _____   _____   _____   _____   _____   _____   8000

4. Write using Roman numerals.

a. 8	b. 19	c. 40	d. 90
12	24	44	76

5. Round these numbers to the nearest hundred.

a.	b.	c.	d.
416 ≈ _____	529 ≈ _____	670 ≈ _____	254 ≈ _____
837 ≈ _____	960 ≈ _____	557 ≈ _____	147 ≈ _____

6. Solve the word problems.

   a. One refrigerator costs $245 and another costs $68 less than that. Find the cost of the cheaper refrigerator. Also, estimate it using rounded numbers.

   My estimate: about $ _____

   b. Mr. Sandman bought two of the cheaper refrigerators, and paid with $400. What was the total cost?

   What was his change?

7. Draw lines using a ruler.

   a. 7 cm 8 mm

   b. 10 cm 5 mm

   c. 2 1/2 inches

   d. 4 3/4 inches

8. Fill in the blanks, using the units in, ft, or mi.

   a. Ann's living room is 20 ____ wide.   **b.** The refrigerator is 28 ____ wide.

   c. It is about 2 ____ to the bookstore.   **d.** The doctor is 6 ____ tall.

9. Fill in the blanks, using the units cm, km, mm, and m.

   a. The fly was 12 ____ long.   **b.** The room measures about 3 ____ .

   c. Mark rode his bike 12 ____ to go home.   **d.** The teddy bear was 25 ____ tall.

Review Chapter 9

1. Write a multiplication and a division fact to match the picture.

a. ____ × ____ = ____

____ ÷ ____ = ____

b. ____ × ____ = ____

____ ÷ ____ = ____

2. Divide.

a.	b.	c.	d.
36 ÷ 6 = ____	44 ÷ 11 = ____	56 ÷ 7 = ____	0 ÷ 9 = ____
3 ÷ 3 = ____	60 ÷ 6 = ____	72 ÷ 9 = ____	16 ÷ 16 = ____
36 ÷ 3 = ____	25 ÷ 5 = ____	99 ÷ 9 = ____	12 ÷ 1 = ____
4 ÷ 1 = ____	54 ÷ 9 = ____	100 ÷ 10 = ____	12 ÷ 2 = ____

3. Make fact families.

a.	b.	c.
____ × 6 = 42	____ × ____ = ____	____ × ____ = ____
____ × ____ = ____	____ × ____ = ____	____ × ____ = ____
____ ÷ ____ = ____	____ ÷ 8 = 1	____ ÷ ____ = ____
____ ÷ ____ = ____	____ ÷ ____ = ____	49 ÷ ____ = 7

4. Find the missing numbers.

a. ____ × 5 = 45	b. ____ ÷ 5 = 4	c. ____ ÷ 3 = 3	d. 72 ÷ ____ = 8
e. 8 × ____ = 96	f. 56 ÷ 8 = ____	g. 54 ÷ ____ = 9	h. ____ ÷ 8 = 8

5. Multiply. Then for each multiplication, write two matching divisions **if possible**. Some divisions are not possible!

a. 6 × 0 = ____ ____ ÷ ____ = ____ ____ ÷ ____ = ____	b. 1 × 9 = ____ ____ ÷ ____ = ____ ____ ÷ ____ = ____	c. 0 × 0 = ____ ____ ÷ ____ = ____ ____ ÷ ____ = ____

6. Divide and find the remainder.

a. 11 ÷ 2 = ___ R ___	b. 41 ÷ 8 = ___ R ___	c. 16 ÷ 5 = ___ R ___
d. 56 ÷ 10 = ___ R ___	e. 26 ÷ 4 = ___ R ___	f. 22 ÷ 9 = ___ R ___

7. Solve the word problems. Write a division or a multiplication for each problem.

a. The teacher bought six boxes of crayons with eight in each box. How many crayons does she have? ____ ☐ ____ = ____	b. The coach of a swimming club put 24 children into groups of six. How many groups did that make? ____ ☐ ____ = ____
c. Rachel packaged cookies in bags to sell them. She had 48 cookies and she put 6 cookies in each bag. How many bags of cookies did she have? ____ ☐ ____ = ____	d. Harry has put 94 stamps in his stamp book with ten stamps on each page. How many pages are full of stamps? ____ ☐ ____ = ____

Puzzle Corner

What numbers can go into the puzzles?

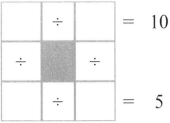

Chapter 10: Fractions
Introduction

The last chapter of *Math Mammoth Grade 3* deals with a few elementary fraction concepts: the concepts of a fraction and of a mixed number, fractions on a number line, equivalent fractions, and comparing fractions.

First, the student learns to identify fractions in visual models, and to draw "pie models" for some common fractions. You can also use manipulatives or the fraction cutouts provided. In the download version, they are found in their separate folder, and in the printed version, they are appended to the answer key.

Next, students represent fractions on a number line diagram by partitioning the interval from 0 to 1 into equal parts. They also study fractions on number lines that go up to 3 and learn to write whole numbers as fractions.

The lesson about mixed numbers relies on visual models and number lines. I strongly feel that students first need to understand fraction operations and concepts with the help of visual models or manipulatives, and not introducing the various rules for calculations too soon. Students match fractions and mixed numbers, and even convert mixed numbers back into fractions using visual models. The actual rule for the conversion is not introduced on this level.

Next, we study equivalent fractions. Students recognize and generate simple equivalent fractions using visual models and number lines.

Lastly, students compare fractions in special cases, such as when they have the same numerator or the same denominator, or when the comparison can be made from visual models. They also learn that comparisons are valid only when the two fractions refer to the same whole.

The Lessons

	page	span
Understanding Fractions	155	*4 pages*
Fractions on a Number Line	159	*4 pages*
Mixed Numbers	163	*4 pages*
Equivalent Fractions	167	*3 pages*
Comparing Fractions 1	170	*3 pages*
Comparing Fractions 2	173	*2 pages*
Mixed Review Chapter 10	175	*2 pages*
Fractions Review	177	*3 pages*

Helpful Resources on the Internet

Use these online resources as you see fit to supplement the main text.

Matching Fractions Level 1
Match each fraction to its visual model.
http://www.sheppardsoftware.com/mathgames/fractions/memory_fractions1.htm

Fractions Splat
Four levels: (1) Identify equal or unequal parts; (2) Identify shapes that are divided into halves, thirds, and fourths; (3) and (4) Find the visual model that matches the given fraction.
http://www.sheppardsoftware.com/mathgames/earlymath/fractions_shoot.htm

Concentration from Illuminations
A matching game you can play by yourself or against a friend, matching fractions to equivalent visual representations. (The game also allows you to play a matching game with whole numbers, shapes, or multiplication facts.) Available also for your phone or tablet.
http://illuminations.nctm.org/Activity.aspx?id=3563

Fraction Frenzy 4
Choose the pizza picture that matches the fraction shown using the four arrow keys.
http://www.mathwarehouse.com/games/our-games/fraction-games/fraction-frenzy-4/

Fractions Matcher
Match each fraction or mixed number with the corresponding picture.
http://phet.colorado.edu/sims/html/fraction-matcher/latest/fraction-matcher_en.html

Fraction Fling
Identify the fractions or mixed numbers that are illustrated by the models by "shooting" them with a slingshot.
http://www.abcya.com/fraction_fling.htm

Puzzle Pics—Number Line Fractions
Drag the puzzle piece to the number line that illustrates the given fraction, and reveal the mystery picture!
http://www.mathplayground.com/puzzle_pics_fractions.html

Animal Rescue: Fractions Number Line Game
Move the arrow to the correct place on the number line and rescue the animals!
http://www.sheppardsoftware.com/mathgames/fractions/AnimalRescueFractionsNumberLineGame.htm

Conceptua Fractions: Identify Fractions
A visual tool that shows fractions or mixed numbers using a pie, a bar, dots, and a number line.
https://www.conceptuamath.com/app/tool/identifying-fractions

Clara Fraction's Ice Cream Shop
Convert improper fractions into mixed numbers and scoop the right amount of ice cream flavors onto the cone.
http://mrnussbaum.com/clarafraction/

EQUIVALENT FRACTIONS

Equivalent Fractions
Construct two other, equivalent fractions to the given fraction using a circle or a square. Use the sliders to divide your shape into a certain amount of parts, then click on the parts to color some of them. Click the check mark to check if you got the equivalent fractions right.
http://illuminations.nctm.org/Activity.aspx?id=3510

Conceptua Math: Equivalent Fractions
In this tool, you can use pie, rectangular, or number line model. Divide each shape into parts using the sliders. Then click on parts to color or uncolor them.
https://www.conceptuamath.com/app/tool/equivalent-fractions

Equivalent Fractions Shoot
Click the fraction picture that is equivalent to the given fraction. Choose "Level 1" for this grade level.
http://www.sheppardsoftware.com/mathgames/fractions/equivalent_fractions_shoot.htm

Laura Candler's Fraction File Cabinet
This web page offers several free printables, activities, and games for grades 3-6.
http://www.lauracandler.com/filecabinet/math/fractions.php

COMPARING

Balloon Pop Fractions
Pop the balloons in order from the smallest to the largest fraction.
http://www.sheppardsoftware.com/mathgames/fractions/Balloons_fractions3.htm

Ordering Fractions
Drag the fractions into the right order, from the lowest to the highest.
http://www.topmarks.co.uk/Flash.aspx?b=maths/fractions

Fractions Gallery - Ordering Simple Fractions
Put the fractions in order, from the least to the greatest.
http://www.free-training-tutorial.com/math-games/fractions-gallery-game.html

Conceptua Math: Order Fractions on a Number Line
First create fractions using the button on the top right, then lock them. Use the "dot" button to see them placed on the number line. Then you can use the buttons on the left to see the fractions represented in different ways. Lastly, drag the fractions under the number line dots, and press the check mark.
https://www.conceptuamath.com/app/tool/place-fractions-on-a-number-line

GENERAL

Fractioncity
Children learn about comparing fractions, equivalent fractions, and addition of fractions while they drive toy cars on the "fraction streets". This is not an online activity but a craft-type activity.
http://www.teachnet.com/lesson/math/fractioncity.html

Understanding Fractions

Fractions are formed when we have a WHOLE that is divided into so many EQUAL parts.	
A whole is divided into *two* equal parts. One part is one half. $\frac{1}{2}$	A whole is divided into *six* equal parts. One part is one sixth. $\frac{1}{6}$
A whole is divided into *ten* equal parts. One part is one tenth. $\frac{1}{10}$	Four parts are colored, and the whole has four equal parts. **Four fourths.** $\frac{4}{4}$
Three parts are colored. There are seven equal parts. **Three sevenths.** $\frac{3}{7}$	Two parts are colored, and the whole has five equal parts. **Two fifths.** $\frac{2}{5}$

$\frac{3}{8}$
"three eighths"

The number ABOVE the line tells **how many parts** we have (the colored parts).

The number BELOW the line tells **how many *equal* parts** the **whole** is **divided** into.

After halves, we use ordinal numbers to name the fractional parts (thirds, fourths, fifths, sixths, sevenths, and so on).

1. Color the parts to illustrate the fraction.

a. $\frac{7}{8}$ b. $\frac{6}{10}$ c. $\frac{4}{6}$ d. $\frac{4}{5}$ e. $\frac{2}{4}$ f. $\frac{4}{7}$

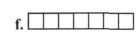

g. $\frac{2}{6}$ h. $\frac{11}{12}$ i. $\frac{5}{9}$ j. $\frac{1}{5}$ k. $\frac{9}{10}$ l. $\frac{2}{7}$

2. Write the fractions, and read them aloud.

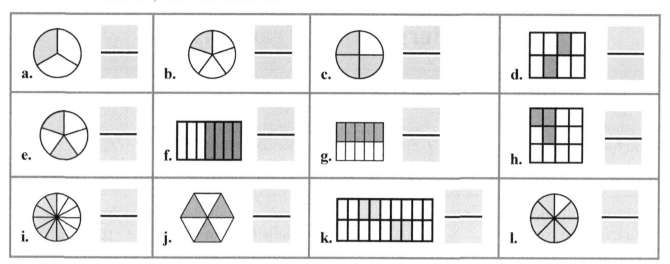

How many parts is this "whole" divided into?
Count. You should get 8 parts.

Don't count the little lines. Count the "units" or the parts. One of them is like this:

How many of them are colored?

You should get 3 colored parts out of 8 in total. So, the fraction is $\frac{3}{8}$.

3. Write the fractions, and read them aloud.

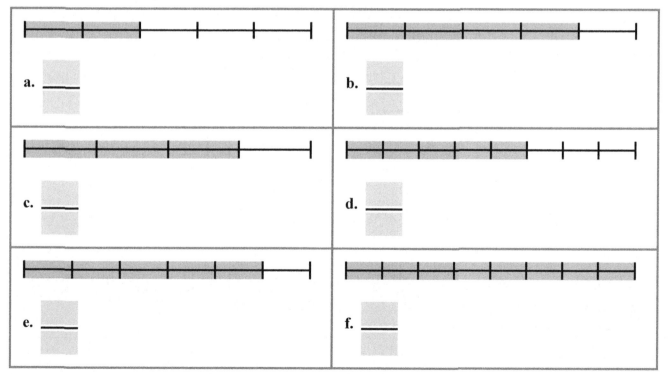

How to draw pie models		
Halves: Split the circle with a straight line.	Thirds: Draw lines at 12 o'clock, 4 o'clock, and 8 o'clock.	Fourths: First draw halves, then split those like a cross pattern.
Fifths: 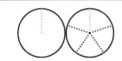 Draw legs and arms like a man doing jumping jacks.	Sixths: First draw thirds, then split those.	Eighths: First draw fourths, then split those.

4. Draw the pie models and color the parts to illustrate the fractions.

a. $\frac{2}{3}$ ◯	b. $\frac{2}{5}$ ◯	c. $\frac{3}{6}$ ◯	d. $\frac{6}{8}$ ◯
e. $\frac{4}{5}$ ◯	f. $\frac{3}{8}$ ◯	g. $\frac{1}{3}$ ◯	h. $\frac{7}{8}$ ◯

5. Color in the whole shape = 1 whole. Then write 1 whole as a fraction.

a. 1 = $\frac{9}{9}$	b. 1 = ——	c. 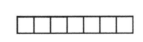 1 = ——	d. 1 = ——	e. 1 = ——

6. Divide the shapes into equal parts and color some of the parts to show the fractions.

a. $\frac{1}{2}$ △	b. $\frac{2}{2}$ ⬡	c. $\frac{1}{3}$ ▭	d. $\frac{3}{4}$ □
e. $\frac{3}{3}$ □	f. $\frac{1}{6}$ ◯	g. $\frac{4}{5}$ ▭	h. $\frac{3}{4}$ ▭

7. Divide the shapes into equal parts. Shade *one* part. Write the area of that part as a fraction of the whole area.

a. Divide the shape into two equal parts.  shaded area = $\dfrac{1}{2}$ of the whole area	**b.** Divide the shape into three equal parts. 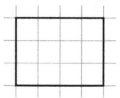 shaded area = ―― of the whole area
c. Divide the shape into six equal parts.  shaded area = ―― of the whole area	**d.** Divide the shape into four equal parts. 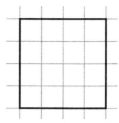 shaded area = ―― of the whole area
e. Divide the shape into three equal parts. 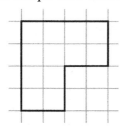 shaded area = ―― of the whole area	**f.** Divide the shape into five equal parts. 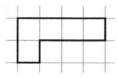 shaded area = ―― of the whole area
g. Divide the shape into four equal parts. 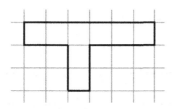 shaded area = ―― of the whole area	**h.** Divide the shape into four equal parts. 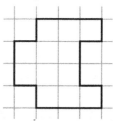 shaded area = ―― of the whole area

Fractions on a Number Line

This is a number line from 0 to 1. It is divided into *six* parts. One part is this much: ⊢─┤. One part is NOT a little line, so do not count the little lines to count the parts.

The arrow marks the fraction $\frac{2}{6}$: two parts are colored, of six parts in total.

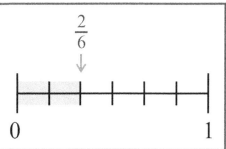

Here you see all the fractions from $\frac{0}{6}$ to $\frac{6}{6}$ on the number line. The fraction $\frac{5}{6}$ is marked with a dot.

Notice that $\frac{0}{6}$ is the same as 0, and $\frac{6}{6}$ is the same as 1.

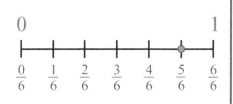

1. Fill in. You can also color a rectangle that goes up to the arrow, to help you.

a. The number line from 0 to 1 is divided into _____ parts.

The arrow marks the fraction ──.

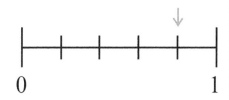

b. The number line from 0 to 1 is divided into _____ parts.

The arrow marks the fraction ──.

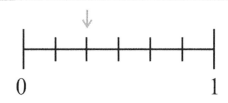

2. Write the fractions under every tick mark.

159

3. Write the fraction shown by the big dot on the number line.

a. 	b. 
c.	d.

4. Mark the fraction on the number line (with a dot).

a. $\dfrac{2}{4}$ 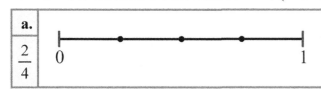	b. $\dfrac{3}{5}$ 
c. $\dfrac{6}{9}$ 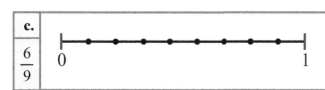	d. $\dfrac{1}{8}$ 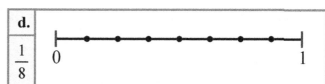

5. Divide the number line from 0 to 1 into equal parts. Then mark the fraction on it.

a. Divide this into two parts. $\dfrac{1}{2}$	b. Divide this into four parts. $\dfrac{1}{4}$ 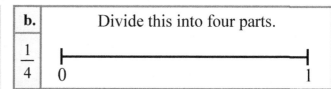
c. $\dfrac{1}{3}$ 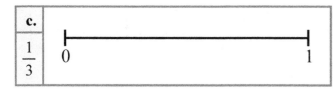	d. $\dfrac{1}{5}$ 
e. $\dfrac{3}{4}$ 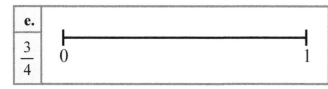	f. $\dfrac{2}{3}$ 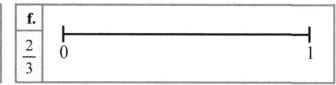
g. $\dfrac{4}{5}$ 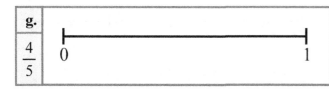	h. $\dfrac{1}{6}$ 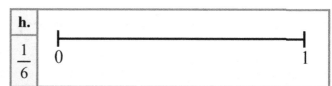
i. $\dfrac{5}{6}$ 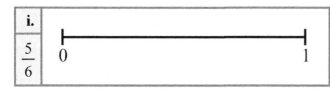	j. $\dfrac{3}{8}$ 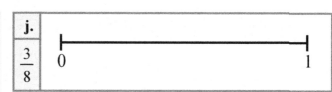

Here, the number line from 0 to 1 is divided into **four** parts. The number line from 1 to 2 is also divided into **four** parts, and similarly from 2 to 3.

So, each little part is one fourth.

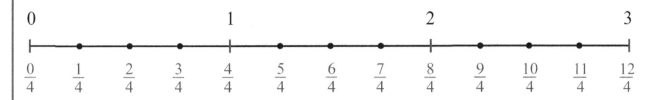

Notice especially: $1 = \frac{4}{4}$, $2 = \frac{8}{4}$, and $3 = \frac{12}{4}$.

In fact, the fraction line is another symbol for division. So, $\frac{8}{4}$ is the division problem $8 \div 4$!
Thinking that way, it is easy to see that $\frac{8}{4} = 2$. You just divide!

6. Write all the fractions under the tick marks.

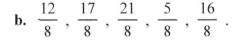

7. Mark the fractions on the number lines. In each list, one fraction is actually a whole number. Which one?

a. $\frac{7}{6}$, $\frac{11}{6}$, $\frac{18}{6}$, $\frac{3}{6}$, $\frac{13}{6}$.

b. $\frac{12}{8}$, $\frac{17}{8}$, $\frac{21}{8}$, $\frac{5}{8}$, $\frac{16}{8}$.

8. Count the parts (thirds, fourths, fifths, sixths, or eighths), and write the whole numbers as fractions.

| a.  1 = ____ | b. 2 = ____ | c. 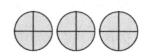 3 = ____ |

Can you find a shortcut so that you don't actually have to *count* all those pie pieces?

| d. 2 = ____ | e.  3 = ____ | f. 1 = ____ |

| g.  4 = ____ | h. 4 = ____ |

9. Divide the pies into parts, and color the pies. Write the whole numbers as fractions.

a. Divide into halves.

 4 = ___/2

b. Divide into eighths.

3 = ___/8

c. Divide into sixths.

 3 = ___/6

d. Divide into fourths.

 4 = ___/4

10. These fractions are actually whole numbers! What numbers are they? (Hint: Divide.)

a. $\frac{6}{6}$ =	b. $\frac{21}{7}$ =	c. $\frac{24}{6}$ =	d. $\frac{20}{2}$ =
e. $\frac{20}{4}$ =	f. $\frac{8}{8}$ =	g. $\frac{12}{3}$ =	h. $\frac{30}{5}$ =

Puzzle Corner Write these whole numbers as fractions.

a. 5 = ___/6   b. 7 = ___/5   c. 3 = ___/7   d. 6 = ___/10   e. 9 = ___/5

Mixed Numbers

Mixed numbers have two parts: • the whole number part, and • the fractional part.	 "One *and* one-third"	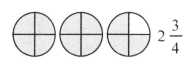 "Two *and* three-fourths"

1. Write what mixed numbers the pictures illustrate. Read the mixed numbers. Remember to use "and" between the whole-number part and the fractional part.

a. _____

b. _____

c. _____

d. _____

e. _____

f. _____

2. Draw pie pictures to illustrate these mixed numbers.

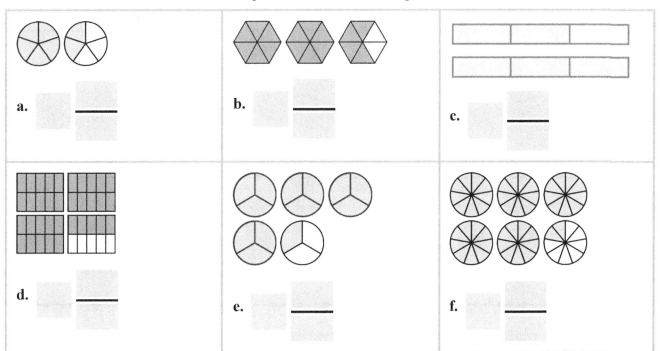

a. $1\frac{1}{2}$

b. $2\frac{2}{3}$

c. $2\frac{3}{5}$

d. $4\frac{5}{6}$

e. $3\frac{5}{8}$

Study this number line carefully.

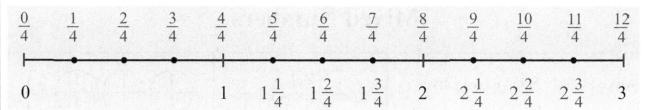

After 1, we have two ways of marking the fractions for the tick marks on the number line:

- either as fractions, such as $\frac{7}{4}$ or $\frac{11}{4}$, counting all the fourth parts from zero.

- or as *mixed numbers*, with a whole-number part and a fractional part, such as $1\frac{3}{4}$ or $2\frac{3}{4}$.

3. Mark the mixed numbers on the number lines.

a. $1\frac{2}{6}$, $2\frac{1}{6}$, $2\frac{5}{6}$

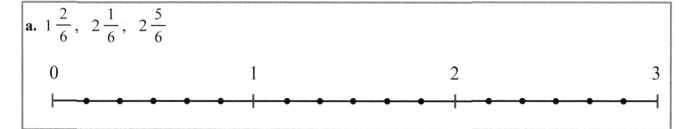

b. $1\frac{5}{8}$, $1\frac{1}{8}$, $2\frac{4}{8}$, $2\frac{6}{8}$

c. $1\frac{1}{5}$, $3\frac{4}{5}$, $2\frac{2}{5}$, $4\frac{3}{5}$

4. Use the number lines above, and write these fractions as mixed numbers.

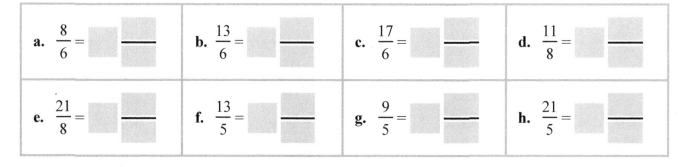

a. $\frac{8}{6} =$ ___

b. $\frac{13}{6} =$ ___

c. $\frac{17}{6} =$ ___

d. $\frac{11}{8} =$ ___

e. $\frac{21}{8} =$ ___

f. $\frac{13}{5} =$ ___

g. $\frac{9}{5} =$ ___

h. $\frac{21}{5} =$ ___

You can use pie pictures to write a mixed number as a fraction.

Just count the fractional parts.

 Count all the thirds. You should get <u>four</u> thirds.

So, $1\frac{1}{3} = \frac{4}{3}$.

 Count all the fourths. You should get <u>eleven</u> fourths.

So, $2\frac{3}{4} = \frac{11}{4}$.

5. Match.

$1\frac{3}{4}$ 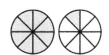 $\frac{9}{4}$

$2\frac{1}{4}$  $\frac{9}{8}$

$1\frac{1}{8}$ 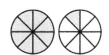 $\frac{12}{5}$

$3\frac{1}{5}$ 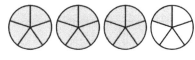 $\frac{11}{8}$

$1\frac{3}{8}$ 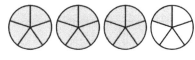 $\frac{7}{4}$

$2\frac{2}{5}$ 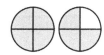 $\frac{16}{5}$

6. Write these both as mixed numbers AND as fractions.

a.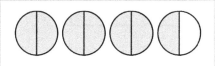
$\dfrac{}{} = \dfrac{}{2}$

b.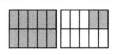
$\dfrac{}{} = \dfrac{}{}$

c.
$\dfrac{}{} = \dfrac{}{}$

d.
$\dfrac{}{} = \dfrac{}{}$

e.
$\dfrac{}{} = \dfrac{}{}$

f.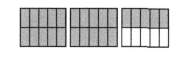
$\dfrac{}{} = \dfrac{}{}$

This is a 1-liter pitcher. In other words, when it is full, it holds 1 liter of water. It is divided into fourths.

Here we have $1\frac{1}{4}$ liters of water.

7. Write a mixed number to tell how much water is in the pitchers.

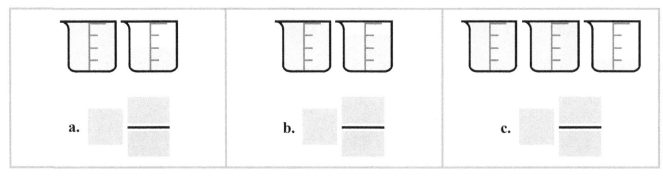

a. _____   b. _____   c. _____

8. Remember measuring lines in inches? Measure the lines using the ruler. Your answers will be *mixed numbers*.

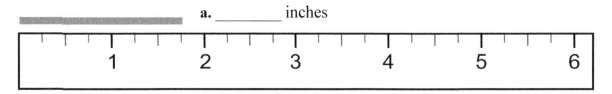

a. _____ inches

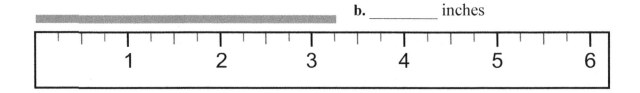

b. _____ inches

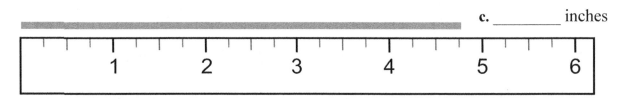

c. _____ inches

9. Draw a line that is...

   a. 4 3/4 inches long

   b. 6 1/4 inches long

Equivalent Fractions

If you eat half of a pizza or 2/4 of a pizza you have eaten the same amount. The two fractions are **equivalent**.

We can write an equal sign between them: $\frac{1}{2} = \frac{2}{4}$.

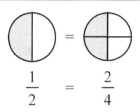

The dot for $\frac{3}{5}$ is in the same place on the number line as the dot for $\frac{6}{10}$. Again, the two fractions are equivalent. We can write $\frac{3}{5} = \frac{6}{10}$.

1. Write the equivalent fractions.

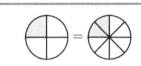

a. ___ = ___

b. ___ = ___

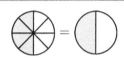

c. ___ = ___

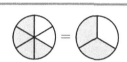

d. ___ = ___

e. ___ = ___

f. ___ = ___

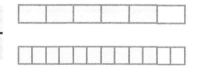

2. Write the equivalent fractions.

a. ___ = ___

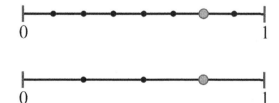

b. ___ = ___

3. Shade the parts for the first fraction. Shade the same *amount* in the second picture. Write the second fraction.

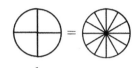
a. $\dfrac{1}{4}$ =

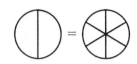
b. $\dfrac{1}{2}$ =

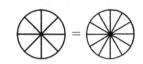
c. $\dfrac{6}{8}$ =

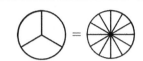
d. $\dfrac{2}{3}$ =

e. $\dfrac{1}{3}$ = ―――

f. $\dfrac{8}{12}$ = ―――

4. Mark the equivalent fractions on the number lines.

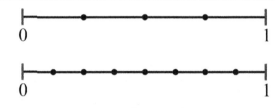
a. $\dfrac{3}{4} = \dfrac{6}{8}$

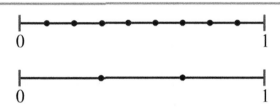
b. $\dfrac{3}{9} = \dfrac{1}{3}$

c. $\dfrac{3}{6}$ = ―――

d. $\dfrac{2}{6}$ = ―――

5. Mark the equivalent fractions on the number lines. This time, you need to first divide each number line into equal parts.

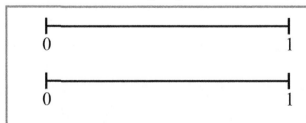
a. $\dfrac{2}{4} = \dfrac{1}{2}$

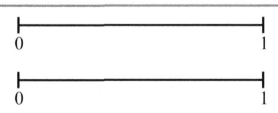
b. $\dfrac{2}{3} = \dfrac{4}{6}$

6. Color and write many fractions that are equivalent to the first fraction.

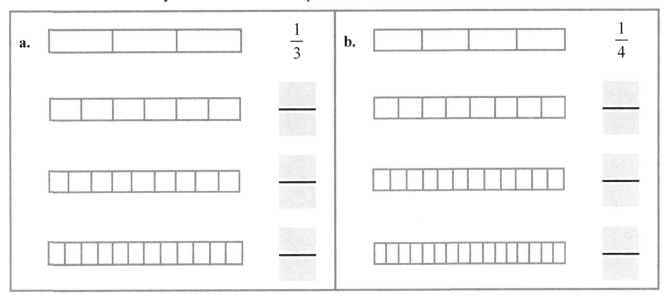

7. Four children have a chocolate bar to share. Cassy says, "Let's divide it into four equal pieces, and everybody gets one piece." Hannah says, "No, let's divide it into twelve equal pieces and everybody gets three pieces."

   Whose idea lets everybody get a fair share?

8. Draw a picture to show that 1/2 = 4/8.

9. **a.** Half of the pie is left. Show in the picture how three persons can share it equally.

   b. What two equivalent fractions can you write from your "cutting"?

10. Are 5/5 and 4/4 equivalent fractions?
    Why or why not?

Puzzle Corner Which is longer, a line that is 3 1/2 inches long or a line that is 3 1/4 inches long?
How much longer is it?

Comparing Fractions 1

1. Color one piece in each "pie." Then compare the fractions. Write < or >.

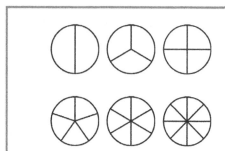

a. $\dfrac{1}{3}$ ☐ $\dfrac{1}{2}$ b. $\dfrac{1}{2}$ ☐ $\dfrac{1}{5}$

c. $\dfrac{1}{5}$ ☐ $\dfrac{1}{4}$ d. $\dfrac{1}{6}$ ☐ $\dfrac{1}{5}$

e. $\dfrac{1}{6}$ ☐ $\dfrac{1}{8}$ f. $\dfrac{1}{2}$ ☐ $\dfrac{1}{8}$

2. Color these fractions in the fraction bars.

   $\dfrac{1}{5}$ $\dfrac{1}{10}$ $\dfrac{1}{2}$ $\dfrac{1}{4}$

   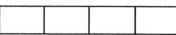
   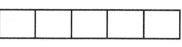
   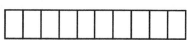

   Find the <u>smallest</u> fraction (the smallest piece).

   Find the <u>greatest</u> fraction (the largest piece).

3. Show, using the two number lines, that $\dfrac{1}{3}$ is greater than $\dfrac{1}{4}$.

   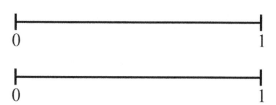

4. Which is a greater fraction, $\dfrac{1}{9}$ or $\dfrac{1}{8}$?

   Explain how you can know that.

5. Write these four fractions in order from smallest to largest: $\dfrac{1}{6}$ $\dfrac{1}{3}$ $\dfrac{1}{9}$ $\dfrac{1}{5}$

   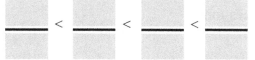 < ___ < ___ < ___

170

6. Compare and write > or < between the fractions. Which is more pie to eat?

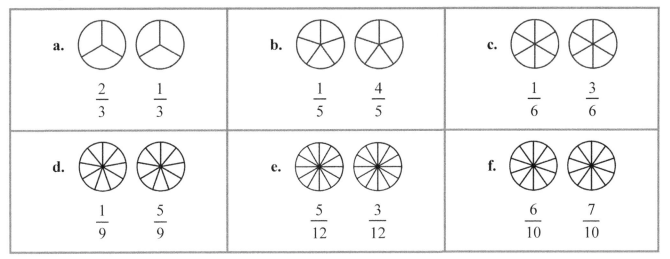

| a. $\frac{2}{3}$ ___ $\frac{1}{3}$ | b. $\frac{1}{5}$ ___ $\frac{4}{5}$ | c. $\frac{1}{6}$ ___ $\frac{3}{6}$ |
| d. $\frac{1}{9}$ ___ $\frac{5}{9}$ | e. $\frac{5}{12}$ ___ $\frac{3}{12}$ | f. $\frac{6}{10}$ ___ $\frac{7}{10}$ |

7. Explain how to find the bigger fraction if two fractions have the same kind of parts (the bottom numbers are the same)? For example, 5/8 and 3/8.

_____

8. Compare and write > or < between the fractions.

a. $\frac{2}{6}$ ___ $\frac{2}{3}$	b. $\frac{2}{5}$ ___ $\frac{2}{8}$	c. $\frac{3}{6}$ ___ $\frac{3}{4}$
d. $\frac{5}{6}$ ___ $\frac{5}{8}$	e. $\frac{2}{2}$ ___ $\frac{2}{3}$	f. $\frac{4}{8}$ ___ $\frac{4}{5}$
g. $\frac{8}{10}$ ___ $\frac{8}{9}$	h. $\frac{3}{8}$ ___ $\frac{3}{6}$	i. $\frac{7}{12}$ ___ $\frac{7}{9}$

9. Explain how to find the bigger fraction if two fractions have the same amount of pieces (the top numbers are the same)? For example, 5/8 and 5/7.

_____

10. Mark the fractions on the number lines. Then use the number lines to compare the fractions. The fraction that is furthest from 0 is the bigger fraction.

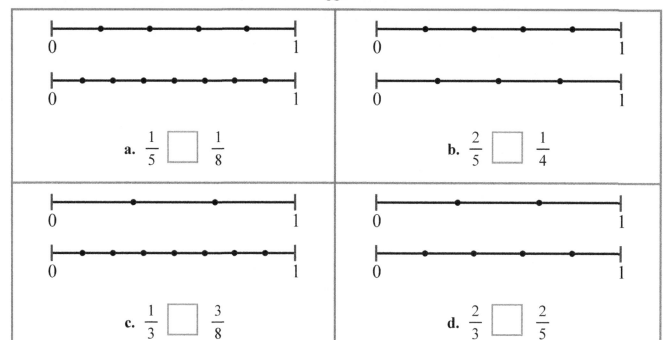

11. Mark the fractions on the number lines. Then find the biggest fraction of the three.

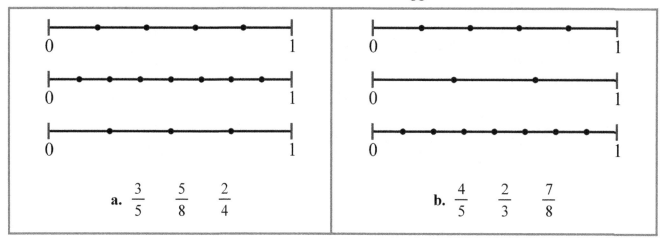

12. Compare the fractions. Think carefully if the fractions have the same amount of pieces or the same kind of pieces. Either way, you can know which one is greater!

a. $\dfrac{5}{7}$ ☐ $\dfrac{4}{7}$ b. $\dfrac{7}{5}$ ☐ $\dfrac{7}{10}$ c. $\dfrac{9}{12}$ ☐ $\dfrac{5}{12}$ d. $\dfrac{10}{20}$ ☐ $\dfrac{10}{12}$

e. $\dfrac{3}{7}$ ☐ $\dfrac{3}{4}$ f. $\dfrac{7}{11}$ ☐ $\dfrac{10}{11}$ g. $\dfrac{5}{8}$ ☐ $\dfrac{5}{6}$ h. $\dfrac{5}{8}$ ☐ $\dfrac{9}{8}$

Comparing Fractions 2

 Both of these paint cans are 1/3 full. But the bigger paint can has more paint than the smaller. In other words, 1/3 of the smaller can is less than 1/3 of the bigger can.

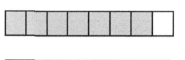 $\frac{7}{8}$ Here, these two fractions look like they are the same amount. Yet, 7/8 cannot be equal to 9/10. Or can it?

 $\frac{9}{10}$ To tell for sure, the wholes that we take a fraction (part) of, have to be the **same size**. In these fraction bars, the total length of each fraction bar is not the same!

Now we have two wholes that are the same size.

Now we can compare, and see that 9/10 is slightly greater than 7/8.

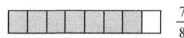 $\frac{7}{8}$

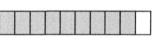 $\frac{9}{10}$

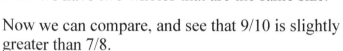

1. **a.** These two pitchers are both 1/4 full. Do they have the same amount of water?

   b. Who got to eat more pie? Can you tell?

    Jess   Mary

2. Compare the fractions. Write >, <, or =. If you cannot compare because the wholes are not the same size, then cross the whole problem out.

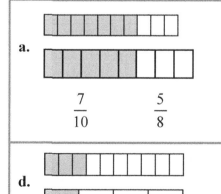

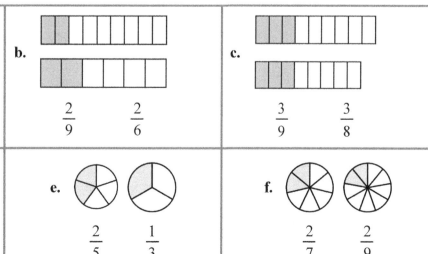

a. $\frac{7}{10}$ $\frac{5}{8}$

b. $\frac{2}{9}$ $\frac{2}{6}$

c. $\frac{3}{9}$ $\frac{3}{8}$

d. $\frac{3}{10}$ $\frac{1}{4}$

e. $\frac{2}{5}$ $\frac{1}{3}$

f. $\frac{2}{7}$ $\frac{2}{9}$

3. Six workers bought one large pizza and one small pizza to share evenly. Both pizzas are cut into three equal pieces. Will it be fair if everybody gets one piece (1/3)?
Why or why not?

4. Janet has two bars of dark chocolate. Which is more, 1/12 of the bigger bar, or 2/12 of the smaller bar?

Does that prove that $\frac{2}{12} < \frac{1}{12}$?

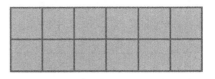

5. Ellie has a blue and a red ribbon that are the same length. She cuts the blue ribbon into 5 equal pieces, and the red ribbon into 4 equal pieces. Which is longer, a piece from the blue ribbon or a piece from the red ribbon?

6. Jack has two paint cans, one big and one small. Both of them are 3/4 full. Which can has more paint in it? (Hint: draw a picture!)

7. Compare the fractions. Think carefully.

   a. $\frac{4}{3} \square \frac{3}{3}$
   b. $\frac{6}{7} \square \frac{6}{9}$
   c. $\frac{9}{10} \square \frac{7}{10}$
   d. $\frac{9}{12} \square \frac{9}{5}$

   e. $\frac{1}{6} \square \frac{1}{4}$
   f. $\frac{1}{12} \square \frac{10}{10}$
   g. $\frac{3}{8} \square \frac{3}{6}$
   h. $\frac{1}{2} \square \frac{8}{8}$

8. Margaret wants to show us that $\frac{1}{5} = \frac{1}{4}$ using the number lines. Your task is to draw a picture to prove to her that it is NOT so!

Mixed Review Chapter 10

1. Divide.

a.	b.	c.	d.
56 ÷ 7 = ____	48 ÷ 6 = ____	54 ÷ 9 = ____	48 ÷ 8 = ____
49 ÷ 7 = ____	72 ÷ 6 = ____	81 ÷ 9 = ____	72 ÷ 8 = ____
28 ÷ 7 = ____	54 ÷ 6 = ____	36 ÷ 9 = ____	32 ÷ 8 = ____

2. Write matching division and multiplication sentences.

a.	b.	c.
____ × ____ = ____	3 × 0 = ____	____ × ____ = ____
42 ÷ 7 = ____	____ ÷ ____ = ____	____ ÷ ____ = ____
____ ÷ ____ = ____	____ ÷ ____ = ____	72 ÷ 8 = ____

3. Divide and show the remainder.

a.	b.	c.
16 ÷ 5 = ____ R ____	21 ÷ 4 = ____ R ____	19 ÷ 6 = ____ R ____
12 ÷ 5 = ____ R ____	27 ÷ 4 = ____ R ____	31 ÷ 6 = ____ R ____

4. Kathy needs to read a 27-page booklet in three days. If she reads the same amount each day, how many pages will she read each day?

5. Six children are sharing 20 apples equally. How many apples will each child get? How many apples will be left over?

6. Write a number sentence for the shaded area and solve.

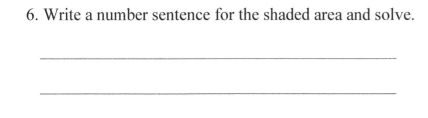

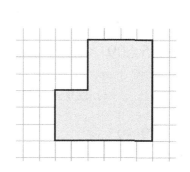

7. Round these numbers to the nearest ten, and
estimate the perimeter of this park.

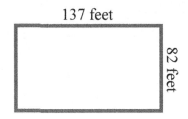

Estimate: _____ ft

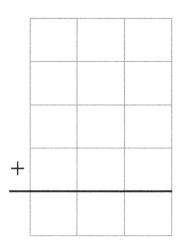

Also find the real perimeter by adding the original numbers in columns.

8. Write the fraction that the arrow points to on the number line.

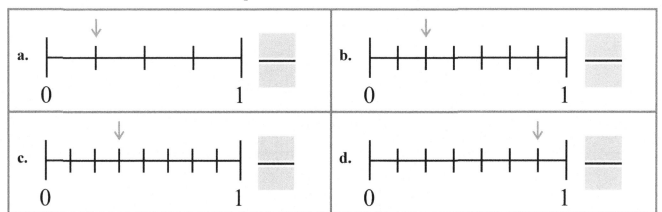

9. Explain how to find which is the greater fraction: $\frac{7}{10}$ or $\frac{7}{8}$?

_____

_____

10. One of the three numbers fits on the empty line so that the comparisons are true.
    Which number? Circle the number.

a. 5,637 5,673 5,607	b. 6,142 6,121 6,211
5,609 < _____ < 5,650	6,114 < _____ < 6,140
c. 6,996 9,966 9,696	d. 4,001 4,010 4,011
9,595 < _____ < 9,700	4,001 < _____ < 4,011

Fractions Review

1. Shade in the fractions. Then compare, and write >, <, or = between them.

   a. $\frac{2}{9}$ $\frac{2}{10}$ b. $\frac{5}{7}$ $\frac{5}{9}$ c. $\frac{1}{4}$ $\frac{1}{3}$

2. Divide the shapes into equal parts. Shade *one* part. Write the area of that part as a fraction of the whole area.

   a. Divide the shape into seven equal parts.

   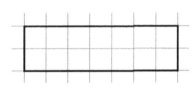

   shaded area = ―― of the whole area

   b. Divide the shape into five equal parts.

   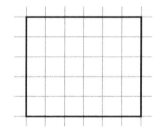

   shaded area = ―― of the whole area

3. Write the fraction shown by the big dot on the number line.

   a.   b.

   c.   d.

4. Mark the fraction on the number line (with a dot).

   a. $\frac{5}{8}$

   b. $\frac{2}{9}$

5. Write all the fractions under the tick marks.

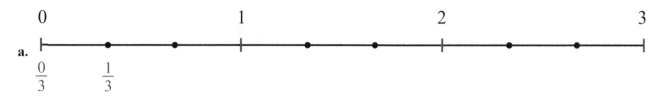

6. Write these fractions as mixed numbers or whole numbers. Use the number line.

a. $\frac{4}{3}$ =	b. $\frac{8}{3}$ =	c. $\frac{6}{3}$ =	d. $\frac{9}{3}$ =

7. Write the whole numbers as fractions.

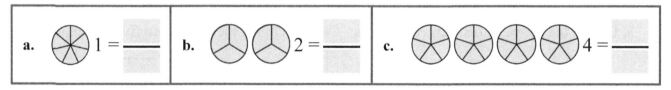

8. Draw pictures to illustrate these mixed numbers.

a. $1\frac{1}{3}$	b. $2\frac{3}{4}$

9. Use the fraction bars on the right to write two fractions that are equivalent to 1/3.

$$\frac{1}{3} = \underline{} = \underline{}$$

10. Write these fractions in order from smallest to largest. You can use the fraction bars to help you.

$$\frac{1}{3} \qquad \frac{3}{6} \qquad \frac{2}{9} \qquad \frac{4}{9}$$

11. Compare the fractions. Write <, > or = between them.

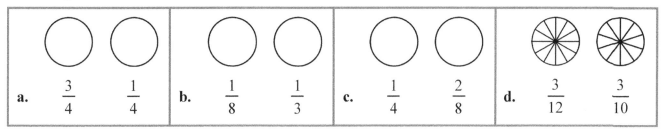

12. Compare the fractions. Write <, > or = between them.

  a. $\dfrac{6}{8}$ ☐ $\dfrac{7}{8}$ b. $\dfrac{1}{5}$ ☐ $\dfrac{1}{10}$ c. $\dfrac{2}{9}$ ☐ $\dfrac{2}{5}$ d. $\dfrac{1}{2}$ ☐ $\dfrac{2}{4}$

13. Explain how to find which is the greater fraction: $\dfrac{8}{9}$ or $\dfrac{5}{9}$?

    _____

14. Mark the equivalent fractions on the number lines.

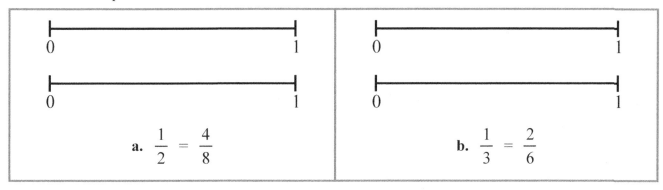

15. Write and shade the equivalent fractions.

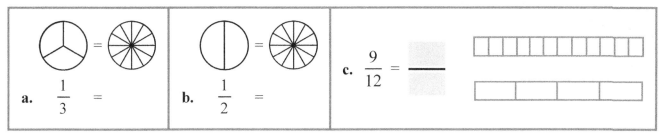

16. Margaret drew these two pictures to show that $\dfrac{2}{8} = \dfrac{2}{4}$.

    What do you think? Is she correct?

Made in the USA
Middletown, DE
05 September 2020